THE HISTORY OF THE STEEL HELMET IN THE FIRST WORLD WAR

Also by Haselgrove and Radovic
Helmets of the First World War: Germany, Britain & their Allies

Also by Radovic
German Helmets of the Second World War (*two volumes*)
German Camouflaged Helmets of the Second World War (*two volumes*)

THE HISTORY OF THE STEEL HELMET IN THE FIRST WORLD WAR

VOLUME TWO

· GREAT BRITAIN · GREECE · HOLLAND ·
· ITALY · JAPAN · POLAND ·
· PORTUGAL · ROMANIA · RUSSIA ·
· SERBIA · TURKEY · UNITED STATES ·

MICHAEL J. HASELGROVE
& BRANISLAV RADOVIC

Schiffer Military History
Atglen, PA

Acknowledgments

The authors wish to thank the following people who have contributed to the preparation and content of this book by providing photographs, information or access to their private collections. We have, wherever possible, given credit to the owners of the photographs or helmets that appear in the book. We are particularly grateful to the following for their generosity in sharing their knowledge and/or precious possessions with us: David Acon, Michael Baldwin, Jean Marc Barris, Alan Beadle, John Burnett, Stephen Chambers, Marcus Cotton, Malcolm Fisher, Daniel Fisher, Brian Maederer, David Powers, Greg Nevill, Tony North, Andy Read, Ellwood von Seibold, Robert Tredwen, Matthew Tredwen, Charles Woolley.

In addition, we would like to thank the following institutions and their staff:

The Imperial War Museum, London and Duxford:
Martin Boswell, Diana Condell, Alan Cant, Tom Duplock, Helena Nicholls

The Polish Army Museum, Warsaw:
Director Jacek Macyszyn, Wojciech Potrzebnicki, Zdzislaw Zajdler

Le Musée Militaire du Perigord, France:
President Jacques Dornier, Jacques Moniotte, Claude-Jean Blanchard

All the above museums are a "must visit" for anyone interested in military history. The authors strongly recommend them to their readers.

The authors would be grateful to any of our readers who might have additional information, material or photographs that could enhance our collective knowledge of the subject and perhaps be included in any update of these volumes.

Lastly, we would like to thank our long-suffering wives for their assistance and great patience: Judy Haselgrove and Anna Chajkin.

Book design and photography by Robert Biondi and Branislav Radovic.

Library of Congress Catalog Number: 2006925465.

Printed in China.
ISBN: 0-7643-2529-9

We are always looking for people to write books on new and related subjects. If you have an idea for a book, please contact us at the address below.

Published by Schiffer Publishing Ltd.
4880 Lower Valley Road
Atglen, PA 19310
Phone: (610) 593-1777
FAX: (610) 593-2002
E-mail: Info@schifferbooks.com.
Visit our web site at: www.schifferbooks.com
Please write for a free catalog.
This book may be purchased from the publisher.
Please include $3.95 postage.
Try your bookstore first.

In Europe, Schiffer books are distributed by:
Bushwood Books
6 Marksbury Ave.
Kew Gardens, Surrey TW9 4JF
England
Phone: 44 (0)20 8392-8585
FAX: 44 (0)20 8392-9876
E-mail: info@bushwoodbooks.co.uk
www.bushwoodbooks.co.uk
Free postage in the UK. Europe: air mail at cost.
Try your bookstore first.

GREAT BRITAIN

CHAPTER EIGHT

GREAT BRITAIN

Great Britain declared war on Germany on the 4th August 1914. The declaration of war was made for various reasons. These included treaty obligations to protect the neutrality of Luxembourg and Belgium whose territory had been invaded by Germany. The declaration of war had the effect of bringing various British territories and dominions into the conflict including Australia, New Zealand, India, South Africa and Canada. These countries contributed troops and materiel to the Allied war effort and all used British helmets. They will therefore be considered in this section rather than separately.

When war was declared the British Army, comprising about 725,000 troops, was relatively small when compared with some other countries, such as Germany and France. Despite involvement in a number of conflicts in the latter part of the 19th century, in common with other nations, the British had no generally available protective helmet. Troops serving abroad in hot climates did use helmets, but these were designed for use against the sun rather than anything else. The need for a suitable ballistically protective helmet was not initially appreciated.

The following represents a brief history of the background developments leading to the introduction of a helmet issued to all troops.

The French Adrian Model 1915 Helmet

In June 1915 a committee was established at British Headquarters in France to investigate and report upon new inventions and suggestions. By that stage, the French had already decided to adopt the "Cerveliere" head protection and the British ordered 1,000 of these for evaluation. However, the order coincided with the introduction and delivery to the French of the first Adrian Model 1915 helmets and was therefore cancelled. Adrian helmets were ordered instead and 495 helmets, fitted with the standard infantry badge, were delivered on the 29th July 1915 and subsequently issued for testing.

The Adrian helmets were evaluated during August 1915 and found to be generally useful. An issue of 50 for each infantry battalion was recommended. At that stage helmets were not thought of as something that would be issued to every soldier. Rather, they were regarded as "trench stores", that is items to be left in the front line for use by the relieving troops.

The War Office Committee met at the beginning of August 1915 to consider various examples of helmets that had been submitted. Their opinion was that Adrian Model 1915 helmet was unsatisfactory. It was felt that it provided insufficient protection and the manufacturing process, involving as it did many separate operations, was unduly complicated. Other designs were also considered, one by an enterprising employee of the Army and Navy Store in London. His name was John Leopold Brodie and it was his design the Committee chose. For more information on the "Cerveliere" and Adrian Helmet see section under France.

Private Purchase Helmets

British officers were required, using an inadequate allowance provided to them and their own resources, to purchase their own equipment. When the war started, private manufacturers and retailers, who provided a plethora of equipment for private sale, saw an opportunity to manufacture and sell helmets. Some of the helmets sold privately were as follows:

The Anglo-French Steel Cap Liner

This form of protection, similar to that mentioned in the section under France, was designed for wear under the standard service cap in the same way as the "Cerveliere". Manufactured from high tensile steel of approximately 1.3mm thickness, it weighed about 0.5kg.

The helmet was manufactured in two sections and was adjustable to fit the head. The example in the collection of the Imperial War Museum, London measures 164mm from side to side and 198mm from front to rear when fully extended. There were two slots in the top of each section of the helmet and a nut and bolt passed through each securing the two sections together. The slots measured approximately 6mm each and when the nut on each bolt was loosened the two sections could be adjusted to fit the wearer's head. The interior of the helmet was painted brown and partially lined with a simple leather headband. The unlined section made use of the cloth cap liner.

In November 1915, the Munitions Inventions Department carried out tests on this protection. A bullet fired from a .45 inch Colt automatic pistol failed to penetrate the steel. However the protection was severely damaged, the joint welds giving way.

This form of protection, judging from the lack of any number of surviving examples, was not widely used and indeed, it is possible that the example held in the collection of the Imperial War Museum, is unique.

The "Acier Trempe" (hardened steel) Helmet

These helmets appeared shortly before the standard French Adrian Model 1915 helmet. Officers purchased them in limited numbers. Details of the helmet are given in the section under France. The helmets provided to the British had no badge and it was common for the owner to attach his unit badge to the front.

The helmet bore a striking resemblance to the standard French Adrian Model 1915 helmet and the two should not be confused. For instance, there is a well-known contemporary photograph of Winston Churchill, then a Major serving with the 6th Royal Scots Fusiliers, wearing an Adrian helmet. That helmet, which has the French infantry badge fixed to the front, appears to be a standard French Adrian Model 1915 helmet and not, as might be expected, a privately purchased helmet.

The Brodie Helmet

Attention is drawn to the comments made in the section "Recent research by Marcus Cotton". That research serves to alter the previously accepted position in respect of the early Brodie helmets as detailed below.

John Leopold Brodie filed a patent application for a helmet and liner in August 1915. That patent was granted and became GB11,803/15. Following rejection of the French Adrian Model 1915 helmet, consideration was given to the Brodie design. There was little that was particularly attractive or innovative about the helmet. Considerable importance was placed on the liner. This was said to keep the head away from the helmet shell with the advantage that considerable indentation of the shell could occur before it came into contact with the head. The patent specification stated:

> *"The lining acts as a buffer to prevent concussion of the brain by external impact and to prevent injury to the frontal, parietal or occipital bones of the cranium by any indentation or penetration of the helmet. This buffer or anti-concussion lining is by means of studs spaced away from the helmet shell so that a free space or air gap is left between the lining and shell, and therefore the helmet may be subjected to a considerable force of impact or even indented or penetrated without injury to the wearer".*

The helmet offered a readily available solution to an urgent requirement. It was cheap and easy to manufacture. Further, it appeared to offer good protection from shrapnel bullets coming from above and from falling shell splinters and other debris. After due consideration, it was decided to test the two separate patterns of the helmet, known respectively as the "Type A" and the "Type B", that had evolved.

The "Type A" Brodie Helmet

This first pattern helmet represented a modification of the shape of the helmet shell shown in Brodie's patent drawings. The dome of the helmet was a little narrower and slightly flattened on top. However, the main difference was the brim. This was wider and more sloping than in the original design. The brim measured approximately 40mm at the front and rear and 50mm at each side. The helmet measured 295mm from side to side and 317mm from front to rear.

The "Type B" Brodie Helmet

This pattern was closer to the original patent specification. When compared with the "Type A" helmet it had a more pronounced dome with steeper sides. This gave a more rounded appearance. The brim was considerably smaller than the "Type A" helmet. The helmet shells appear not to have been manufactured uniformly. On one example examined the brim measured approximately 25mm at the front and 29mm at the rear and 35mm at one side and 32mm at the other. The measurement from side to side was 262mm and 288mm from front to rear. On another example the measurements were 259mm from side to side and 287mm from front to rear.

Initially, on the 11th August 1915, the War Office placed an order for fifty "Type A" helmets for experimental purposes. Another order, for one hundred "Type B" helmets, was also placed. Following delivery, at the end of September 1915 forty-nine "Type A" and one hundred "Type B" helmets were dispatched to France for testing. The helmets were manufactured from mild steel approximately .90mm thick. Approval was given by the Commander-in-Chief of British forces for the adoption of the "Type A" helmet. An issue of fifty helmets per battalion was considered appropriate.

On the 24th September 1915, an order was placed for 25,000 "Type A" helmets to be manufactured. This initial order was, it is thought, placed with the Army and Navy Co-operative Society Limited. By the 2nd October 1915, 557 helmets had been produced in addition to the helmets ordered for experimental purposes. On the 12th October 1915 a further order for 10,000 helmets was placed. These helmets were to be produced from non-magnetic manganese steel. Thereafter, all helmets, including those not already manufactured from the previous order, were to be made from this harder, non-magnetic steel. The new steel was found to have approximately three times the resistance of the earlier mild steel used.

The Brodie Helmet War Office Pattern

The new manganese steel, being considerably harder than the mild steel, quickly wore out the dies used to produce the "Type A" helmets. New dies were produced, and the opportunity was taken to slightly alter the shape of the helmet bowl by making it more rounded. The size of the helmet brim was reduced to 32mm at the front and rear and 43mm at the sides. The angle of the brim was also reduced. The helmet shell measured 305mm front to rear and 293mm side to side, although considerable variations in these measurements are possible. The helmet was known as the War Office Pattern.

The Liners

The liners of the "Type A" and "Type B" helmets were in accordance with John Brodie's patent. They consisted of padding fixed to the crown of the helmet made up from layers of felt, lint and cotton wool. This provided a cushion between the wearer's head and the top of the helmet. The liner band, attached to the padding in the crown, supported the one piece, six-tongued liner. The liner was secured at the apex by a drawstring of cotton tape.

Around the outer edge of the headband were fixed, at regular intervals, twelve pieces of tubular rubber measuring approximately 12.5mm wide and 25mm long. These provided a cushion between the head and the side of the helmet. The liner was described in the (retrospectively published) List of Changes dated the 22nd October 1918 as:

> *"Principally of canvas American cloth and buckram. A narrow lining of cordite cloth in which is inserted a row of half-inch by one inch tubular rubbers which serves as a ventilating course and buffer is also provided"*

The liner was held in place by means of a copper rivet. This passed through a hole in the apex of the helmet shell. The liner on the War Office Pattern helmets was modified. This involved changing the material into which the rubber buffers were inserted into a band of leather.

It should be remembered that on many early helmets the original liners were replaced with the later liner used on the Mark 1 helmets. This "refurbishment" extended, in some cases, to replacement of the chinstrap attachment brackets and repainting the exterior of the helmet shell and underside of the brim.

The Chinstrap

The chinstrap was of brown leather approximately 16mm wide and 2mm thick. It was of two-piece construction and was fitted with a pronged buckle to secure it and for adjustment. At each side, the chinstrap was attached to the helmet by passing it through a rectangular wire loop and then fixing with a split-pin. The wire loop, measured externally, was approximately 28mm long and 9mm wide. Made from 2mm wire, the chinstrap attachment loop was fixed to the helmet by means of a rectangular piece of metal folded around the wire loop, with the double thickness of the metal then being attached to the underside of the helmet brim.

On some but not all "Type A" and "Type B" helmets, the attachment was by means of a rivet. On the others, and on later helmets, a split-pin was used.

Paint Finishes

The original "Type A" and "Type B" helmets were given a camouflage paint finish of mottled light green, blue and orange. However, subsequently manufactured helmets, including the War Office Pattern helmets, were painted inside and out in colours varying from a bright apple green to a dull blue-grey. Helmets for use in hot climates were, it appears, painted a yellowish brown colour. By February 1916, the predominant colour used was a dull green. Many helmets were re-painted during the War because the early paint finishes were found to be reflective.

Helmet Production

Production of the new helmets was delayed by a lack of adequate supplies of manganese steel. Indeed, consideration was given to reverting to the use of mild steel. That suggestion was rejected, although in January 1916 a decision was reached to reduce slightly the ballistic resistance requirement. This permitted steel previously rejected as inadequate to be used.

By the beginning of March 1916, 270,000 helmets had been produced. Of these, 140,000 went to the Western Front, with the remainder going to other theatres of war. Helmets were still regarded as "trench stores", the supply being inadequate to provide every soldier with a helmet of his own.

A large-scale test under battle conditions was necessary and an attack scheduled for the 2nd March 1916 was suitable for that purpose. That attack was required to re-capture part of a narrow ridge, thirty to forty feet high, known as "The Bluff", on the northern bank of the Ypres-Comines Canal that had been captured by the Germans on the 14th February 1915. The British attack was a success. A Reuter message from British Headquarters on the 6th March 1916 reported:

> *"The new British casque has had its battle-baptism, and the results of the test have been more than satisfactory... In the assault upon the Bluff, on March 2nd last, and the subsequent fighting, it met with the first great test".*

Certainly it was reported by the Medical Services that severe head wounds were reduced by 75%. Sir Douglas Haig requested that 660,000 helmets be provided for the troops then in France, with 15,000 additional helmets to be produced for each new Division arriving in France. New orders were placed bringing the total orders to that time to 1,255,000 helmets. Of those, 1,000,000 had been supplied by July 1916.

Recent Research by Marcus Cotton

Marcus Cotton has kindly informed the authors that his recent research indicates the following:

- The helmet previously described as "Type B" is more properly referred to as the "1st Pattern Brodie helmet".
- The helmet previously referred to as the "Type A" is, in reality, a version of the "War Office Pattern Helmet". The helmets were produced from an order for "Special 1mm Steel" sufficient to produce 20,000 helmet shells and to be supplied by J.H. Robinson & Co.
- There is a "2nd Pattern Brodie Helmet". This pattern was a leather-covered version of the "1st Pattern Brodie Helmet". Presumably only a very limited number of this pattern were produced. Those that were, probably had the leather covering removed. There are no known examples of this type surviving intact.
- The rubber cushions in the liner were initially solid, thereafter tubular rubber was used.

The above information calls into question some of what has been said previously.

The Helmet, Steel, Mark I

During the testing of the new helmet at the beginning of March 1916 under battle conditions, a number of problems were noted. Summarised, these were:

1. The helmet was too round and shallow making it difficult to keep in place, even using the chinstrap.

2. The liner was hot and slippery.

3. The edge of the helmet brim was sharp and could cause injury.

4. The paint finish was shiny and reflected light.

5. The helmet should be made in at least three sizes.

Following discussions, it was agreed that a new pattern of liner and chinstrap would be introduced. The helmets would be fitted with a metal rim around the brim and would be painted with a non-reflective textured paint. It was agreed that 1,000 helmets of the improved design would be sent to France for evaluation. These helmets were duly sent on the 2nd May 1916 and were approved for service on the 15th May 1916. At that time, a request was made for an extra large size to be provided. Formal approval of the helmet was given in early July 1916 and in September 1916 it became "The Helmet, Steel, Mark I" (hereafter referred to as the Mark I).

By June 1916, most personnel serving overseas received a helmet on arrival. From September 1916, it was possible to issue helmets to troops in training or otherwise engaged at home.

The differences between the Mark I and the War Office Pattern Helmet were as follows:

1. The Mark I liner differed in that it was in two parts. The first consisted of an oval crown pad. This was constructed in layers of lint, asbestos and felt. The pad was attached to the crown of the helmet shell by means of a copper rivet passing through the apex of the helmet. On the base of the rivet in the interior was a nickel-plated washer. This served to hold the pad in place. The rivet also secured in place the top section of the chinstrap assembly. Attached to each side of the upper chinstrap assembly by a split-pin was the second part, namely the headband. That was not dissimilar to the earlier design, having twelve tubular rubber cushions set in a basil leather crown. The headband was padded with cotton wool covered with lint and canvas. Attached to the headband by stitching was the leather-cloth liner. Instead of tongues, this had netting attached to it which was adjustable to fit the top of the head by means of a drawstring.

2. Although the Mark I helmet shell was manufactured in one size only, the new liner came in four sizes. The size was marked normally on the upper part of the chinstrap assembly.

3. In the apex of the Mark I helmet, beneath the washer, a paper label was fitted. This read: *"Tighten Cord and Adjust Net to Fit the Head"*.

4. The Mark I chinstrap was in two parts. The top section, secured at the apex of the helmet by the rivet, came down each side of the helmet bowl and passed through the metal rectangular loop attached to each side of the helmet brim. The lower section of the chinstrap was attached to the upper section on each side by means of a rectangular wire loop.

5. The Mark I chinstrap was fitted with a nickel-plated sliding buckle without a prong instead of the buckle with a prong used on the earlier helmets.

6. The edge of the Mark I helmet brim was fitted with a mild steel rim. This overlapped at the rear of the helmet, usually by 25mm or more, and was fixed by brazing.

7. The wire loops on either side of the Mark I helmet, to which the earlier chinstrap sections had been attached, were enlarged from approximately 10mm to 15mm wide. They remained at approximately 28mm long on both helmets. The enlargement allowed space for the new chinstrap to pass through, rather than being attached to, the rectangular loop.

8. The Mark I helmets were painted in a matte khaki paint inside and out. To the exterior was added sand to provide a textured, non-reflective finish.

9. The Mark I liner cloth was stamped on the underside with a rectangular stamp in red ink giving the patent registration details. The stamp read: *"Brodie's Steel Helmet Registration No 652,800 War Office Pattern. Patent No 11803/16".*

A further modification involved the insertion of a rubber ring, or "doughnut", in the crown of the helmet. This arose following receipt on the 10th January 1917 of a suggestion for that modification from Lieutenant Andrews of the Australian Imperial Forces. The suggestion put forward was on the basis that the rubber ring would provide additional space, or at least a minimum space, between the top of the head and the helmet shell. This would allow for indentation, or even piercing, of the helmet shell without direct injury to the wearer. The force of a blow would, to some extent, be cushioned and spread over an increased area, thus dissipating the force.

The proposal was accepted and in May 1917 General Headquarters requested that all future helmets be fitted with a new pattern crown pad with the rubber ring stitched inside.

Attachment of extra felt to the inner side of the lining below the adjustable netting was a further modification. Additionally, for the basil leather section of the headband, cordite cloth was used instead.

The Manufacture of the Brodie Helmet

All helmet shells were manufactured from manganese steel, apart from the first few thousand helmets produced, and bore the initials or trade mark of the steel supplier and a number to indicate the batch of steel from which the helmet was manufactured. The following tables, taken from an article *"British Steel Helmets 1915-1918"* by Marcus Cotton, may be of interest:

STEEL SUPPLIERS

MANUFACTURER'S CODE	STEEL SUPPLIER	SUPPLIED FROM - TO
FS	Thomas Firth & Sons (Sheffield)	Sep 1915-1919
HS	Hadfield Ltd (Sheffield)	Jan 1916-1919
BS	W. Beardmore & Co Ltd (Glasgow)	Jan 1916-1918
M & S?	J.J. Maxfield & Sons (Sheffield)	Jan 1916-1918?
V	Vickers Ltd (Sheffield)*	1916-1917?
?	Samual Osborn (Sheffield)*	1916-1918?
Imperial	Edgar Allen (Sheffield)*	1916-1918?
?	Bury's & Co (Sheffield)*	1916-1918?
MLS, MS	Miris Ltd (Sheffield)	Mar 1916-1917
JD & S	James Dixon & Sons (Sheffield)	1916-1918?

* Known to have been supplying helmet manufacturers organised by the Sheffield Munitions Committee.

The following markings appear to have been used by the helmet manufacturers organised by the Sheffield Munitions Committee and indicate both the manufacturer and the steel supplier:

D/F	Dixon/Firth
HH/O	Harrison Bros & Howson/Osborn
HV	Huttons/Vickers
R/O	Rounds/Osborn
JD&S	James Dixon & Sons

HELMET MANUFACTURERS

NAME OF MANUFACTURER	PLACE	MANUFACTURED FROM - TO
Army & Navy Co-operative Society Ltd	London	Sep 1915-Jan 1916
Joseph Sankey & Sons Ltd.	Bilston	Oct 1915-1918
Bleriot Ltd.	London	May 1916-Oct 1916
J & J Maxfield & Sons	Sheffield*	Dec 1915-1918?
J Dixon & Sons Ltd.	Sheffield*	Dec 1915-1918?
W & E Viener	Sheffield*	Dec 1915-1918?
Harrison Bros. & Howson	Sheffield*	Dec 1915-1918?
John Round & Sons	Sheffield	1916
W. Hutton & Co. Ltd.	Sheffield*	Dec 1915-1918?
Hadfields	Sheffield**	Aug 1916-1919
Thomas Firth & Sons Ltd.	Sheffield**	Aug 1916-1919
W. Beardmore & Co. Ltd.	Glasgow**	Aug 1916-1918
Miris Steel Co.	Sheffield**	Sep 1916-1917

* Helmet manufacturers organised by the Sheffield Munitions Committee. Each manufacturer only produced small numbers of helmets per week. By 1917 only one or two of the firms were actually manufacturing helmets.

** Prior to August 1916, these firms had supplied steel only.

Only two manufacturers of helmet linings have been identified. These were the Army and Navy Co-operative Society Limited of London and Walsall who were the main manufacturers. However, towards the end of the war, E. Day Ltd of St. Albans, Hertfordshire, who had previously made hats, also contributed to production.

The manufacture of helmets continued until February 1919. Approximately seven million helmets were manufactured. Some of these were provided to allies, including the United States of America.

Cloth Covers

Introduced early, cloth covers were initially to cover the reflective paint finish of the early helmets. However, even after the introduction of textured, non-reflective paint, the extensive use of cloth covers continued. The Calais Ordnance Depot produced covers of sacking, cloth or whatever material was available and, of course, many soldiers made their own. Officers' privately produced versions were usually of four-piece construction made from khaki coloured material with an adjustable drawstring to secure the cover to the helmet

The Use of Insignia and Badges on Helmets

At the start of the Great War, all ranks wore a regimental badge on their service dress caps. Most officers wore regimental collar patches and other ranks wore shoulder titles for their particular regiment.

This system of regimental marking did not identify the higher unit to which the wearer belonged. Markings on vehicles, flags, signposts etc. identified the division. Once the army began to expand, the system was insecure and inadequate.

In addition, although regimental badges were fixed to helmets and helmet covers or were painted or stencilled on, it was obviously not prudent to wear such insignia in the front line as it provided a ready source of information to the enemy. Therefore, in 1916 each division was ordered to select a divisional symbol or sign and subsequently their use became widespread. They served not only as a means of easy identification from a distance, but to conceal the identity from those who did not know the sign. Divisional signs were routinely attached or painted onto the front, side or rear of the helmet, or helmet cover, as well as being worn on the uniform. Fixing the sign to the rear of the helmet, or uniform jacket, assisted in identification for those observing from the rear.

In addition to divisional signs, unit identification "flashes" or "patches" were developed to indicate the regiment of the wearer. These consisted of a shape in a particular colour or pattern of material. Officers might wear a triangular patch, whilst other ranks in the same regiment might use a rectangular patch of the same colour or design. The "badge" or "flash" was either painted onto the helmet or worn as a piece of cloth attached to the helmet cover. In some cases, regimental colours were painted as a band around the base of the helmet bowl. In addition, a rectangular bar in a particular colour was used beneath the "flash" or "badge" to indicate the particular company of the wearer.

Further, insignia or signs were introduced for particular actions or operations with their use being discontinued thereafter. Such badges or signs assisted the wearers in distinguishing friend from foe and helped observers whose function it was to report the progress of a particular operation.

With the creation of specialist units, such as machine-gun companies or trench mortar batteries, other badges were adopted and may be found fixed or painted onto a helmet or cover.

Finally, battle badges were sometimes awarded as recognition of the excellent conduct of an individual or unit.

The wearing of badges and insignia on helmets and helmet covers was widespread and accepted and, in some cases, actively encouraged. However, when fixing a badge, it was inappropriate to cause damage to the helmet shell. On the 14th December 1916, a General Routine Order (GRO 1997) was issued stating:

> *"Drilling holes in steel helmets to enable badges to be worn weakens the structure of its steel and reduces its power of protection. The practice is therefore forbidden. Badges, if required, can be painted or stencilled in a neutral tint".*

In the circumstances, by the end of the War, there existed a complex system of regimental badges, divisional signs, unit signs and battle badges. The use of divisional signs and regimental "flashes" to conceal the identity of the wearer cannot be regarded as a serious proposition. For instance, by 1918, practically all the helmets worn by the 12th (Eastern) Division had stencilled onto the back the divisional sign, the ace of spades, with the regimental badge appearing on the front.

Following the end of the War, the use of the badges developed during the war was gradually abolished. By 1922, the regular army had reverted to the use of regimental badges and rank insignia only, although some Territorial Army units did manage to retain their divisional signs. These were the exception rather than the rule.

Private Purchase "Brodie Style" Helmets

By mid-1916 manufacturers had produced for private sale helmets broadly resembling the shape of the new Brodie helmet. There were several variations in shape and manufacture. One basic design had a fluted bowl. These were possibly manufactured in Birmingham and were of one and two-piece construction, in the case of the latter the helmet bowl being pressed separately from the brim. The two sections were then joined. The Portuguese later adopted this design, although the Portuguese helmets were stamped from a single piece of steel. A second design was similar to the first save that the helmet bowl was not fluted.

A common feature of these helmets was the method of fixing the liner. The attachment was by means of twine, or something similar. This passed through a series of holes punched in pairs and evenly spaced around the base of the helmet bowl and attached to the liner. On some helmets a thin leather strap covered the exterior base of the helmet bowl. The strap appears to serve no useful purpose although it is possible that it was intended to protect the twine. Additionally, a common feature was a layer of thin corrugated cork fixed between the headband of the liner and the helmet shell. The leather headband was, on some helmets, marked with the name and address of the retailer.

The main objection to these helmets was that they were made from thin mild steel. No proper testing was carried out and, in reality, the helmets offered little protection. Ultimately, the sale of the helmets was proscribed.

Private Purchase Brodie Helmets

These helmets used the standard Brodie shell, but had a special liner and chinstrap. Many private retailers sold these helmets. In most cases, the entire helmet was sold. However, it was possible to take a standard issue helmet to a retailer for a private purchase liner and chinstrap to be fitted.

The manufacturers supplying government contracts also supplied the shells used for private purchase helmets. They were therefore of good quality. The liners were usually of high quality and were generally an improvement on the standard design given, in particular, that they were better fitted to the head making the helmet less likely to tilt forwards or backwards when in use. Again, the liners were usually marked with the name and address of the retailer.

One of the best-known retailers was Hawkes & Co of 1, Saville Row, London. Their helmets had a high quality liner initially designed for use in sun helmets. The fitting of these liners involved drilling holes in the helmet bowl. Later liners were designed specifically for steel helmets. Chinstraps were often lined with velvet for additional comfort.

Privately purchased helmets were usually supplied with a cloth cover. The majority of these were constructed from four sections of khaki drill material. However, many variations existed.

In addition, some retailers sold carrying cases. These were intended to be used either by mounted troops when they were attached by means of a strap to the saddle. A shoulder strap was used for those on foot. Few of these carrying cases have survived. Those observed are made of brown leather.

Eye and Face Protection

In late 1915, the Munitions Inventions Department of the Ministry of Munitions considered various designs of facial and eye protection put forward by civilian inventors. The aim was to prevent injuries to the face and, in particular the eyes, caused by small, low velocity shell splinters and other debris.

A committee, formed in June 1916, investigated the issue. One of the members of that committee was Captain Richard R. Cruise, an eminent opthalmic optician. Although experiments had taken place with steel and triplex glass eye protections, he felt these unsatisfactory. Instead, he devised a form of eye protection, designed for attachment to the standard steel helmet.

The Directors of Medical Services in France and the Commander-in-Chief were of the opinion that the Cruise form of eye protection was satisfactory and should be fitted to all helmets. On the 9th September 1916, an order for the manufacture of the Cruise defence was approved.

The defence had a closely woven chain mail face-piece formed from small rings of hardened steel wire and attached by means of twelve larger rings to a steel wire rod or rail. The rail was welded to the underside of the helmet brim. On the sides each end of the wire rail was set into a loop. The loops also served to secure the chinstrap loops. At the bottom of the chain mail visor in each corner was a metal chain. This attached to sliding hooks on the chinstrap. By adjusting these chains it was possible to adjust the fit of the visor. The chinstrap was worn to the rear of the head when the visor was in use. When not in use, the visor could be drawn up and over the helmet brim and secured, using the adjustable chains, to hooks fixed to the upper brim on each side of the helmet.

On the 15th September 1916, an order for 100,000 visors was placed with the Army and Navy Co-operative Society Limited in Walsall, Birmingham. A further order for 250,000 visors was placed on the 28th September 1916. The first delivery, made to Le Havre in France, was on the 7th October 1916.

However, production problems arose. The welding of the metal rail to the underside of the helmet brim caused bad burning of the helmet shell. A revised design, suggested by Captain Cruise, was approved. It was decided to designate the new helmet with visor The Helmet, Steel, Mark II. The helmet, without visor, was to be the Helmet, Steel, Mark I.

The new design involved moving the metal rail supporting the chain mail further forwards and spot welding it to the rim in front of the retaining lugs used to secure the ends of the wire rail and chinstrap. Further problems arose with production in that the chain mail was very time-consuming to manufacture. Production was also extended to the Army and Navy Co-operative Society's factory in London.

In order to fit the new visors to helmets already issued, a new system was devised whereby the metal rail supporting the visor was attached to a formed metal plate. That was, in turn, attached to the helmet brim by means of the same rivet on each side of the helmet securing the chinstrap loops. Samples of this new design were sent to France. On the 30th December 1916, the design was reported as satisfactory. Guidance was issued as to the conversion of Mark I helmets already in use.

On the 16th February 1917, a decision was taken to increase the number of visors required to 500,000. It was decided to ship the visors to France to enable helmets already issued to be converted. By the 24th March 1917, some 112,722 visors had been produced. However, on the 19th May 1917, a report was received from France to the effect that the visor was not now considered so good as the triplex glass goggles that had been considered earlier. Reports on the visor highlighted a number of problems. These were:

1. The visor added to the weight that had to be carried on the wearer's head.
2. Vision was obscured.
3. The loose ends rattled.
4. The balance of the helmet was upset by the extra weight in front.
5. The visor was apt to make men dizzy.

The Commander-in-Chief was not prepared to order men to wear the visor. In the circumstances, contracts for the production of the visors were terminated. By that stage, some 229,876 visors had been delivered.

To overcome the criticisms of the original visor, Captain Cruise designed an improved version. This used steel springs at the sides instead of the chain. In addition, it had a brass stiffener. This was fitted so as to be immediately in front of the wearer's nose, providing more rigidity to the chain mail. The brass stiffener was pivoted, enabling the visor to be folded up and backwards so as to rest on top of the brim and front of the helmet. It was held in place when not in use by the springs on each side. The chain mail manufactured for the earlier design could be used for the new design, thus resulting in a saving.

An experimental batch of 2,000 of the new design was approved in July 1918. The helmets were delivered to France in September 1918 but, due to the end of the war, a decision was taken not to proceed with further production.

The majority of the helmets fitted with the 1st Pattern Cruise Protection were dismantled. However, the dismantled (defrocked) helmets remain easily identifiable because they often retain the remnants of the fittings.

The Tank Crew Protective Helmet

On the introduction of tanks, it was quickly realised that some form of protective headwear was required. When rough ground was crossed, injury resulted to the crew by their coming forcefully into contact with the interior of the tank. Indeed, there was at least one recorded incident when a crew were all knocked unconscious when their tank lurched over a particularly steep obstacle during a demonstration.

Additionally, protection was required against small projectiles that would ricochet around the inside of the tank. These projectiles were often pieces of bullets, or bullet splash, that had penetrated through a gap in the armour where a hole was required for vision or a gun, for instance. Alternatively, they were pieces of metal that had flaked off the armour plate from the inside following a projectile striking the exterior of the tank.

The wide brim of the standard helmet made it unsuitable. A helmet, constructed from four pieces of leather, was accepted for testing. The leather pieces, cut to an appropriate size, were riveted together. Two pieces of leather, running from front to rear, formed the crown of the helmet. There was, in addition, one piece of leather fitted to each side to protect the sides of the head. The helmet had a leather liner and chinstrap.

Orders were placed for the helmet during the summer of 1916. The helmets were available for use on the 16th September 1916, when tanks were used operationally for the first time at Flers-Courcellette in France. The helmets provided adequate protection but it was felt they resembled too closely the German helmet. Tank crews, once outside their tank, might be mistaken for the enemy.

In the circumstances, the helmets were withdrawn from service after a comparatively short time. Total production of the helmet probably reached only a few hundred. Following this, tank crews received standard steel helmets.

The Tank Crew Face Mask

The mask did not become available until November 1917. It consisted of two metal plates. The front plate had a shaped nose piece soldered to it. The rear plate, made from manganese steel, was provided with five slits, or slots, of varying length to allow for vision. The front and rear sections were riveted together. Running between the two sections on either side of the nosepiece, beneath the eye slits, was a piece of wire. This wire was bent outwards to form loops. Each of these passed through one of four holes on each side of the mask. Attached to each of these eight loops thus formed was a metal ring. From these rings, the chain mail face guard was suspended.

The chain mail extended a short distance on each side of the face. The extensions were each supported by a short chain. The chains comprised five links of the same rings used to support the front of the chain mail mask. Each supporting chain was attached directly to the chain mail at the bottom. At the top, each was attached to the bracket on each side used to attach the head tapes.

The rear of the mask was covered with a double thickness of chamois leather with cotton wool between the two layers to act as padding. The front of the mask was covered with a single layer of brown leather. The front and rear coverings were stitched together.

On each side of the mask to the front was riveted a small bracket to which was attached a rectangular metal loop. Two cotton material tapes on each side of the mask, which were tied at the back of the head to secure the mask to the face, were attached to these loops.

The first 1,200 masks were delivered to France on the 29th September 1917. The total number of masks ordered by the end of April 1918 was 22,000. In that year, some 2,000 masks were supplied to the French. American tank crews were also supplied with the masks. In addition, it appears the Germans made use of captured masks.

It is perhaps worth mentioning that modern recreations or illustrations of the mask being used with the leather tank helmet appear inaccurate, given that the leather helmet was withdrawn from service some time prior to the introduction of the face mask.

The masks remained in service until 1927, when stocks were disposed of.

Lightweight Parade Helmets

Very limited use appears to have been made of lightweight parade helmets, pressed from aluminium instead of steel. The only examples noted are of private purchase.

French-made Acier Trempe (hardened steel) private purchase helmet. This example was used by a member of the Motor Machine Gun Corps. Note that the manufacturer's mark is just visible on the neck guard. (Collection of Brana Radovic)

MMG

An Adrian Model 1915 Helmet worn by a British general officer. (Collection of the Imperial War Museum, London)

An Adrian Model 1915 Helmet worn by a member of the Royal Artillery. (Collection of Brana Radovic)

The Anglo-French Steel Cap Liner. (Collection of the Imperial War Museum, London)

A private purchase Brodie-style mild steel helmet with ribbed dome and carrying the cap badge of the West Riding Regiment. The liner is marked: Henry Heath Limited 105-107 Oxford Street London. (Collection of Brana Radovic)

A variation of the private purchase Brodie-style mild steel helmet with ribbed dome. This example has a leather strap and buckle around the helmet bowl, somewhat mimicking a gentleman's hat or tropical sun helmet. (Collection of Brana Radovic)

The same model of private purchase Brodie-style helmet as the previous page, but with a cloth cover fitted. (Collection of Brana Radovic)

The approved 1st Pattern Brodie Helmet. Note the camouflage painted finish and also the two Brodie helmet labels on the liner and the remains of another, larger label on the left side of the helmet. This example is probably the actual prototype with the label that although now rather obscured, reads "Steel Helmet Type B approved 26th September 1915". The letter "B" on the label stands for Brodie and may well account for the previous incorrect description of the "Type B" helmet. It is likely that this helmet is one of the original order of 150 helmets, since only 149 of those were sent to France. (Collection of the Imperial War Museum, London)

BRODIE'S STEEL HELMET
WAR OFFICE PATTERN
5705

Another example of the 1st Pattern Brodie helmet, previously described as the "Type B Brodie" helmet. (Collection of Marcus Cotton)

■ The modified "War Office Pattern" helmet shell, made using the Robinson supplied "special 1mm steel". This helmet was previously described as the "Type A Brodie" helmet. (Collection of Marcus Cotton)

LT. O. D. WILES

A Hadfield steel helmet blank and pressed helmet shell. (Collection of the Imperial War Museum, London)

HADFIELDS LTD SHEFFIELD
18974
TRADE RESISTA 85 MARK

A ballistically tested helmet shell. (Collection of the Imperial War Museum, London)

865 F.S.
790 F.S.

Brodie Helmet "War Office Pattern" (Collection of Brana Radovic). On this and following pages note variations in colour.

Brodie Helmet "War Office Pattern". (Collection of Stephen Chambers)

Brodie Helmet "War Office Pattern". (Collection of the Imperial War Museum, London)

Brodie Helmet "War Office Pattern". (Collection of Marcus Cotton)

Brodie Helmet "War Office Pattern". (Collection of Marcus Cotton)

Brodie Helmet "War Office Pattern". Note the thick textured paste added to prevent reflection. (Collection of Brana Radovic)

Interior view of the shell of a Brodie Helmet "War Office Pattern". (Collection of Brana Radovic)

Helmet Steel Mark I with the textured finish and a modified liner, which is similar in design to the Mark I liner. Note also the paper Brodie Helmet label. (Collection of the Imperial War Museum, London)

IE'S
Registered
WAR
PATENT No

Brodie Helmet "War Office Pattern" shell with textured finish. This helmet has been given a new liner by the Army Ordnance Corps and bears a paper label in the liner crown that reads: AOC Salvage Supply Depot, indicating that it was picked up on the battlefield and re-furbished.

"Helmet Steel Mark I". Note the new chinstrap and liner, now produced in four sizes, the added metal rim, the matte/textured paint finish and the label giving fitting instructions. (Collection of the Imperial War Museum, London)

and Adjust Net to
fit the Head.

Comparisons of helmet shell. Top left: Brodie 1st pattern. Top right: Modified "War Office Pattern". Bottom left: "War Office Pattern". Bottom right: Mark I.

Comparisons of helmet shell. Top left: Brodie 1st pattern. Top right: Modified "War Office Pattern". Bottom left: "War Office Pattern". Bottom right: Mark I.

Comparisons of liner fittings. Top: Brodie 1st pattern. Bottom: Brodie 1st pattern, with the remains of a private purchase liner.

Top: "War Office Pattern".

Bottom: Mark I.

Two-piece hessian drawstring cover. (Collection of Stephen Chambers)

Four-piece hessian drawstring cover. (Collection of Ellwood von Seibold)

One-piece hessian drawstring cover with an added strip with loops for foliage. Note that it is fitted to a Brodie Helmet "War Office Pattern" with its liner turned inside out, presumably for comfort. (Collection of Brana Radovic)

Another example of a one-piece hessian drawstring cover. (Collection of Marcus Cotton)

Two more examples of hessian drawstring covers. Note the different types of material. (Left: Collection of Imperial War Museum, London; Right: Collection of Michael Haselgrove)

Sealed Pattern (official sample for all subsequent production) example of the four-piece tropical drawstring cover for troops in Salonika. (Collection of Imperial War Museum, London)

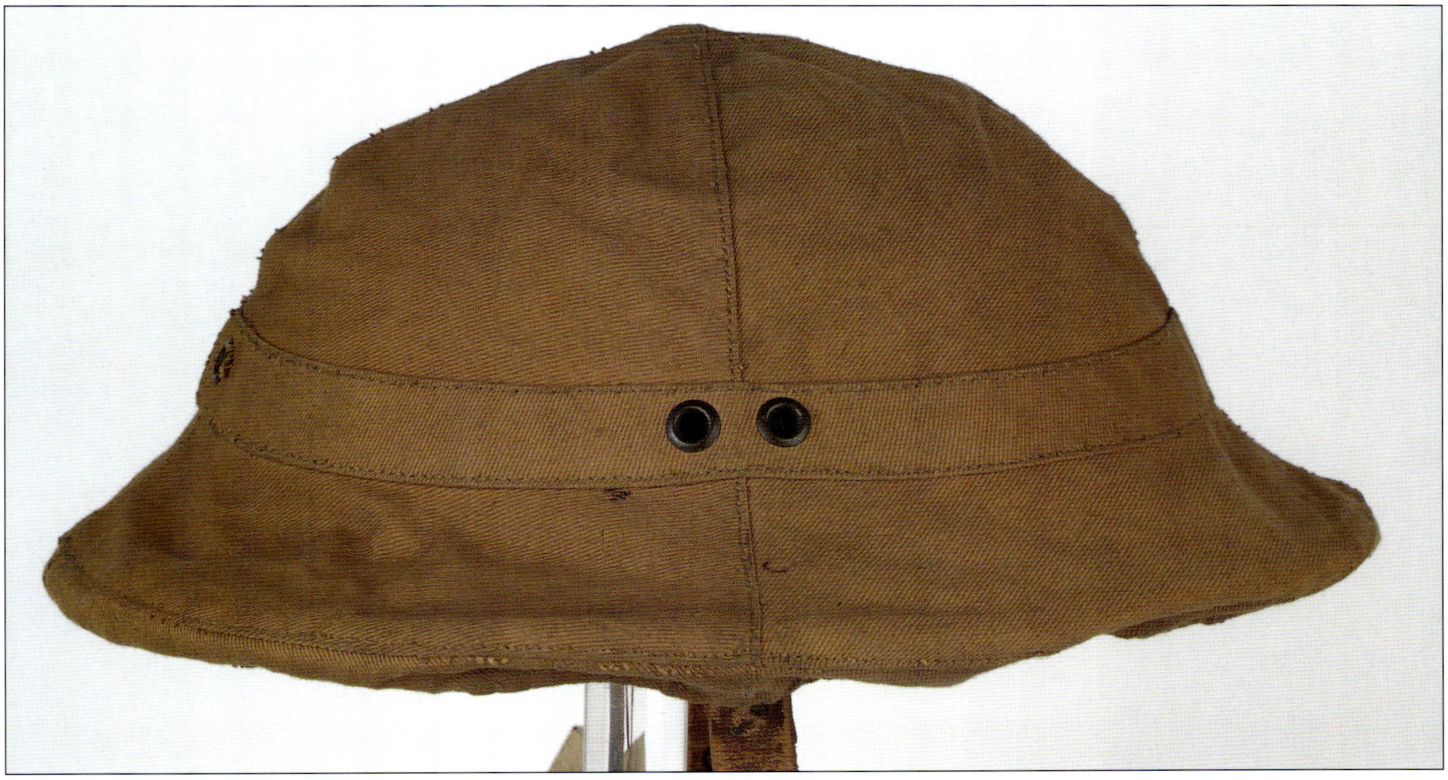

Supplies to be of one Size only as per Sample

Tropical helmet neck flap, which was fitted to the tropical cover. (Collection of Imperial War Museum, London)

A variation of the tropical cloth cover and neck flap. Marked as "Made by Hampton & Sons Ltd, cover by Klinger 1917". (Collection of Stephen Chambers)

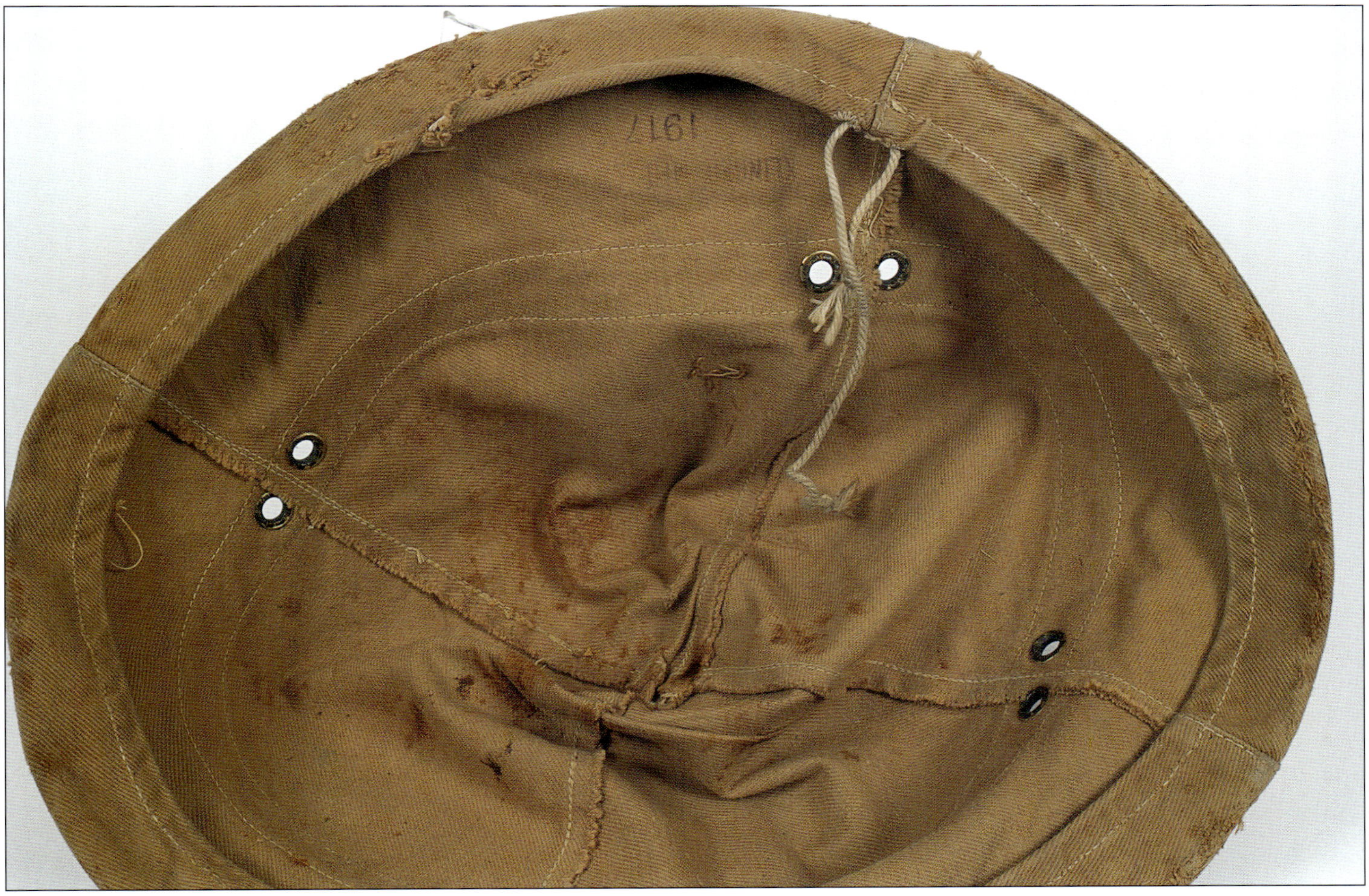

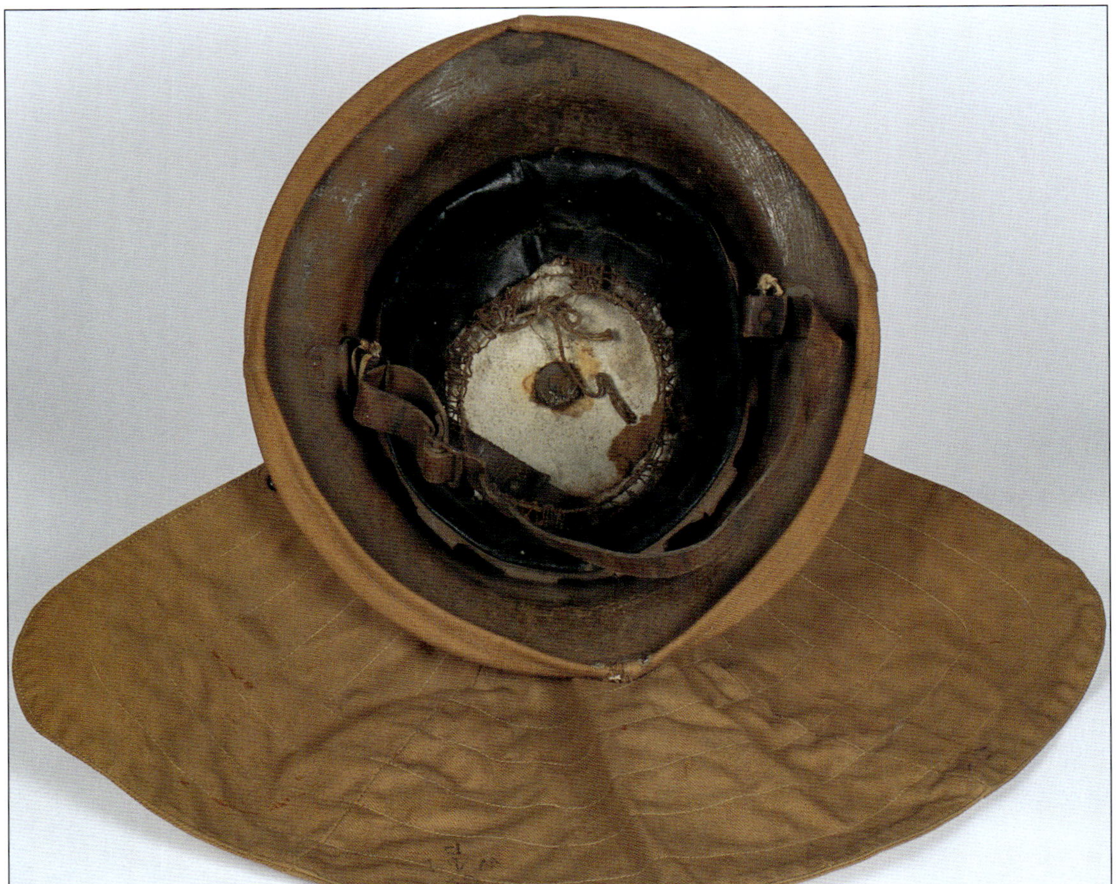

A four-piece hessian cover with the badge of the Royal Berkshire Regiment stencilled across the front. This example is fitted to a "War Office Pattern" helmet and is named to: Corporal Donald H. Bailey (11384, 5th Battalion) and Private Douglas James Rice (10508 who died of wounds 21/9/17). This is a clear example of the re-issuing of helmets, or of wounded soldiers passing helmets onto comrades. (Collection of Brana Radovic)

A two-piece hessian cover with embroidered badge of the London Regiment. The badge is probably the type produced for wear on the side of the issue tropical helmet. (Collection of Brana Radovic)

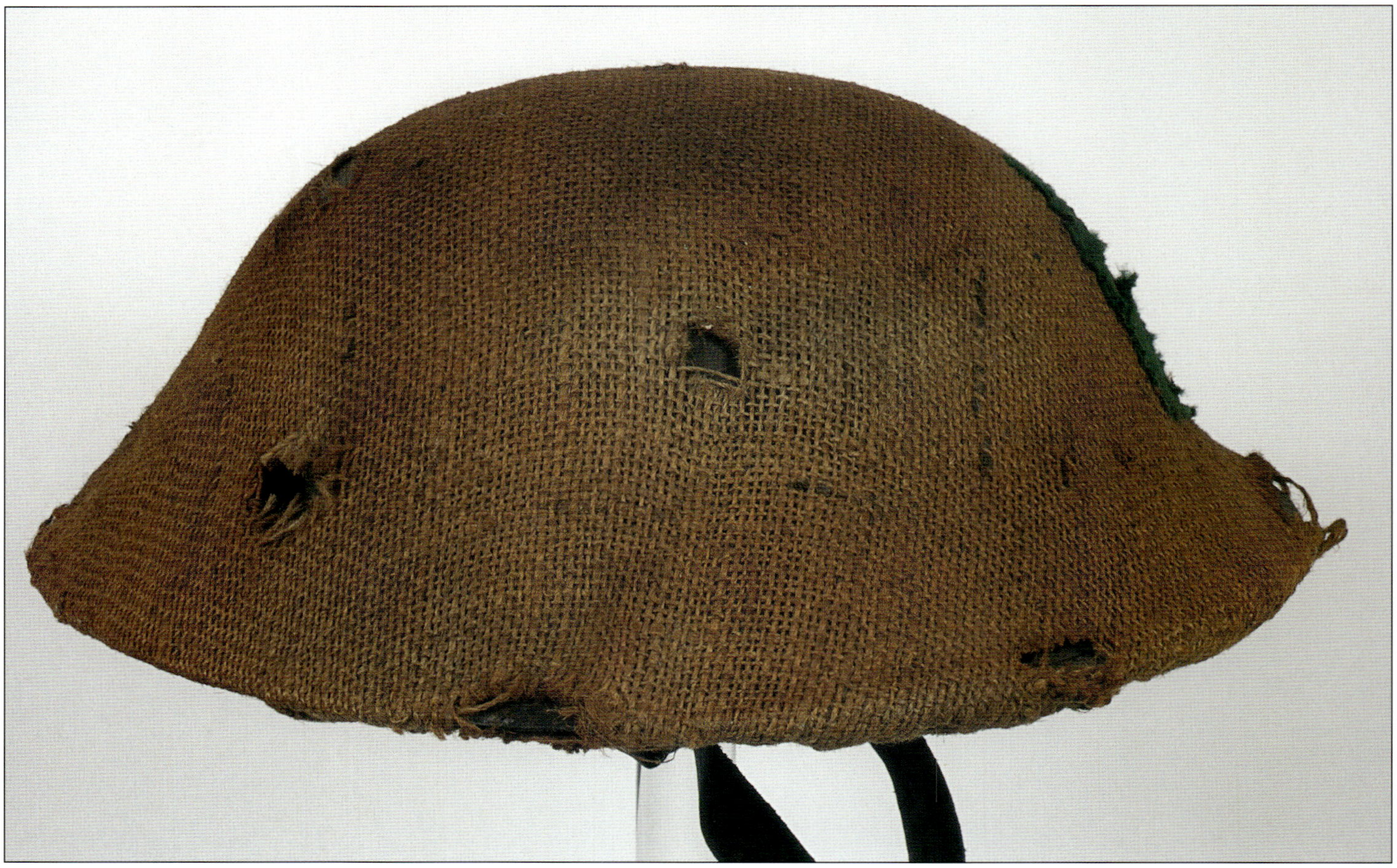

One-piece hessian drawstring cover with the green cloth tactical sign for the Royal Irish Rifles. (Collection of Stephen Chambers)

A variant cloth drawstring cover with a green diamond-shaped tactical sign sewn onto both sides. (Collection of Marcus Cotton)

Officers' private purchase four-piece cloth drawstring cover with a two-piece cloth band sewn around the bowl. (Collection of Brana Radovic)

Officers' private purchase four-piece cloth cover. Note the private purchase officers' liner (this particular example is made by Hawkes, Saville Row, London) and that the chinstrap is lined with velvet for added comfort. This helmet is named to E. Wade (Capt.) Hampshire Regiment. (Collection of Marcus Cotton)

HAWKES'S PATENT SELF FITTING LINING
British 17486, Canadian
India 442, All of 1912.

Another example of the type illustrated on the previous pages. This example was owned and worn by Major G.C. Wheeler V.C. He won the Victoria Cross when serving with the Ghurka Rifles in Mesopotamia in 1917. The liner is a modified Hawkes type. (Collection of Brana Radovic)

A four-piece officers' cover with cloth band attached. The embroidered badge is that of the Royal Naval Division. (Collection of Stephen Chambers)

A two-piece cloth cover that appears to have been made from rubber-lined trench coat material. It carries a General officers' cap badge and was worn by Brigadier General Orr-Ewing, who at one time commanded the 2nd Battalion, The Scots Guards. (Collection of Brana Radovic)

Two-piece officers' cloth drawstring cover with band. It has a metal Royal Engineers' cap badge attached. This helmet is named to Lt. Cyril Richmond Tobitt. (Collection of the Imperial War Museum, London)

Two-piece officers' cloth drawstring cover with a metal Royal Artillery (Territorial) badge attached. Note the application of dark painted camouflage scheme. (Collection of Brana Radovic)

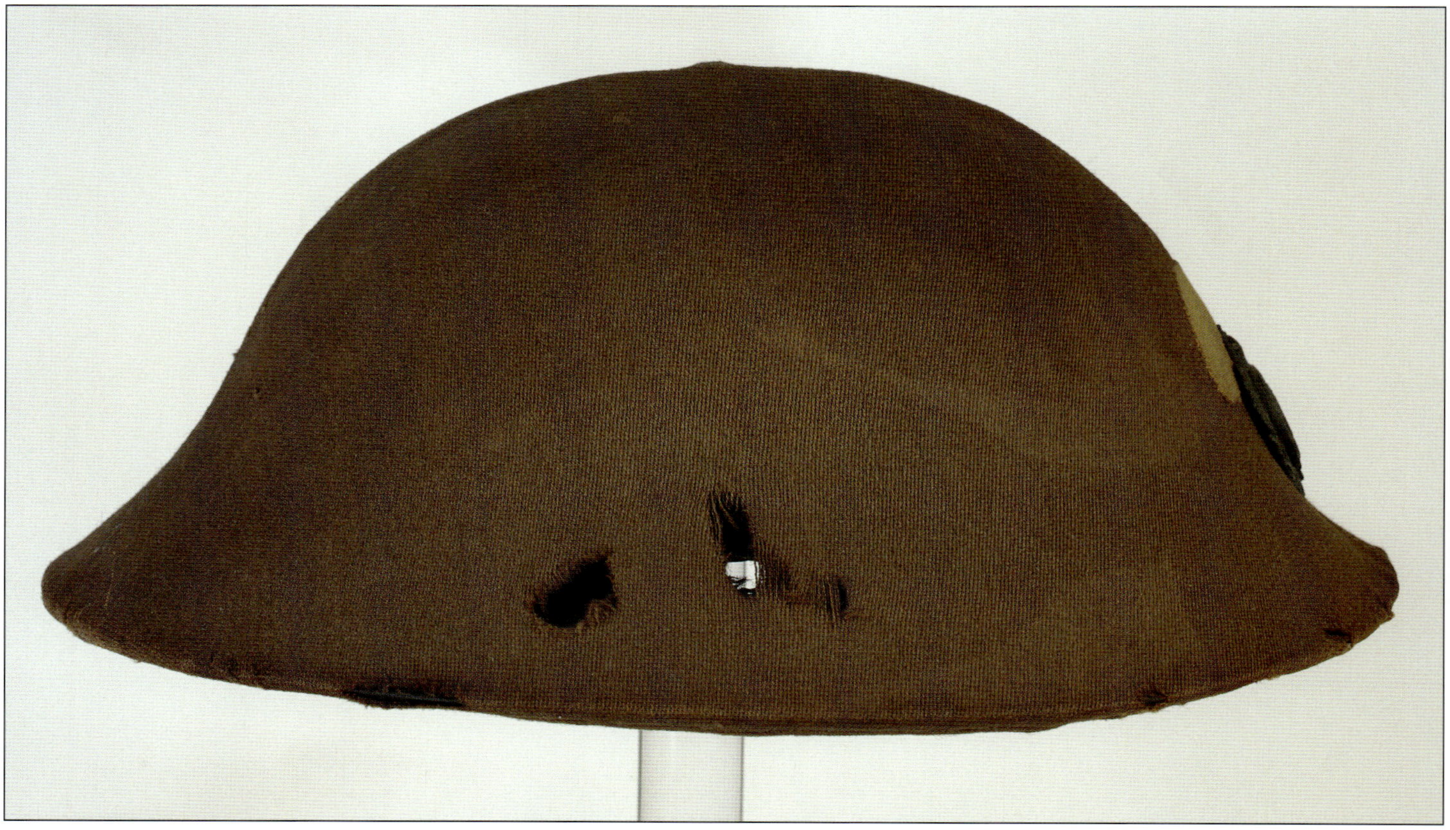

Two-piece officers' cloth drawstring cover with a metal Royal Engineers' badge attached. Note the yellow tactical felt patch above the cap badge. This helmet was hit and pierced by shell splinters, wounding the wearer. (Collection of Brana Radovic)

Two-piece officers' cloth drawstring cover with large metal Guards Regiment badge attached. (Collection of Robert Tredwen)

A private purchase lightweight Brodie-style helmet with embroidered General officers' cap badge attached. It is named to Major General Ramsey (See period photo section) and was made by Belgian helmet and hat makers Jules Fonson. It is possible that this helmet is of immediate post-war manufacture as the main Fonson factory was based in Brussels, which was under German occupation throughout the war. (Collection of Brana Radovic)

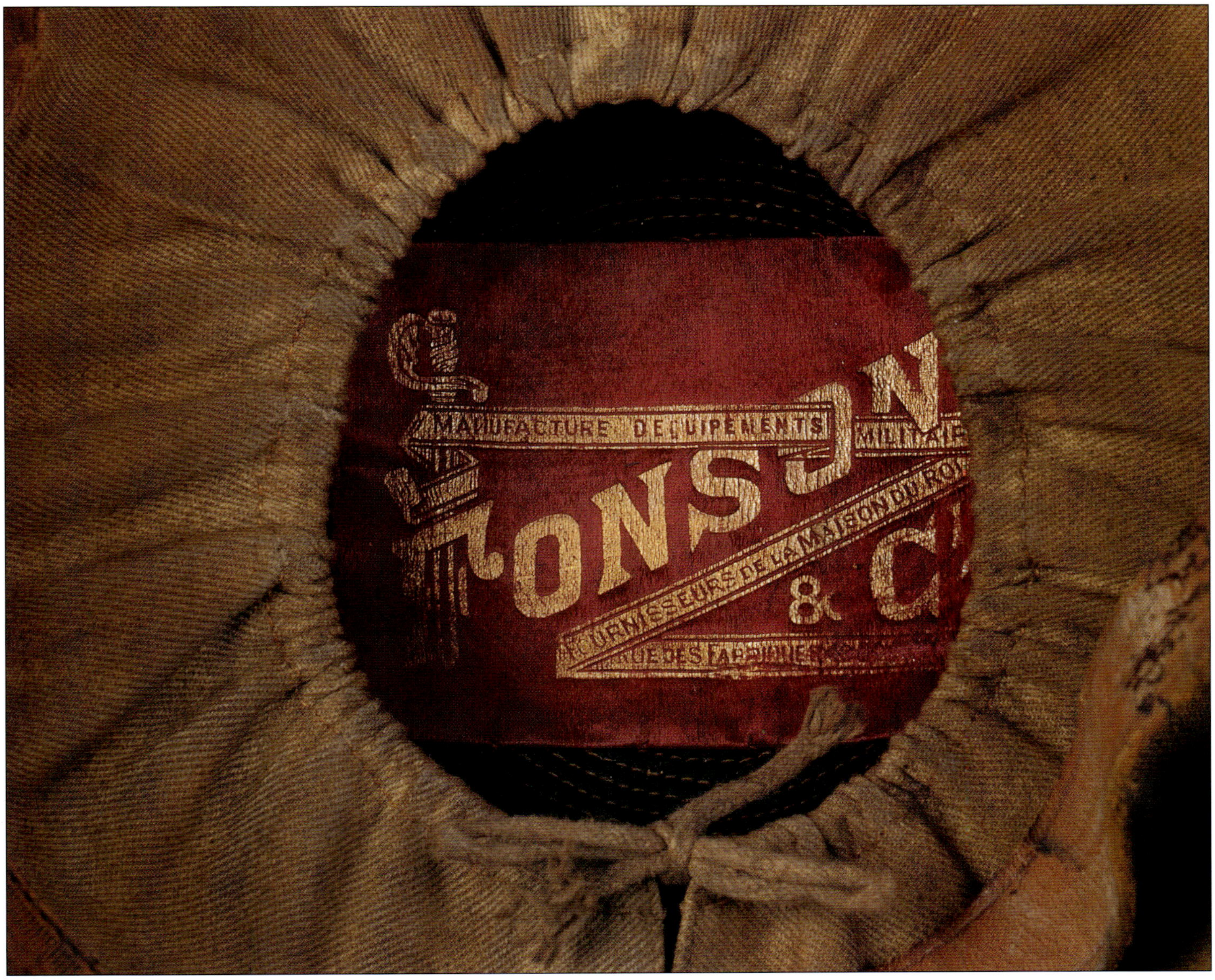

Eight examples of private purchase liners. Note that the one shown at bottom left is the same splinter-damaged Royal Engineers' helmet with yellow tactical badge illustrated on a previous page. Note the blood-stained field dressing and one of the pieces of shell splinter found in the liner. (Collection of Brana Radovic)

THOMAS & STONE
HAT & CAP MAKERS
ST JAMES'S
LONDON, S.W.
A.T. Godley

Helmet, Steel, Mark II (1st version Cruise visored helmet) with Machine Gun Corps insignia stencilled onto the front. (Collection of the Imperial War Museum, London)

Helmet, Steel, Mark II (1st version Cruise visored helmet). Top right shows how the chain mail curtain could be drawn aside to allow a rifle to be sighted. Note the variations of mail fixings on these and following pages. With the rail attached directly to the helmet shell being the early version and the plate arrangement being the later. (Collection of the Imperial War Museum, London)

Helmet, Steel, Mark II (1st version Cruise visored helmet). (Collection of the Imperial War Museum, London)

■ Helmet, Steel, Mark II (1st version Cruise visored helmet). (Collection of Marcus Cotton)

Helmet, Steel, Mark II (2nd version Cruise visored helmet). Improved second pattern helmet with chain mail visor. (Collection of Marcus Cotton)

Examples of de-frocked Helmet, Steel, Mark II (1st version Cruise visored helmets). Top right shows an example with the hooks for holding the veil back removed. Bottom left carries the insignia of the Machine Gun Corps and the 57th division symbol. (Malcolm Fisher, Regimentals)

De-frocked Helmet, Steel, Mark II (1st version Cruise visored helmet) interiors.

■ Examples of painted insignia (front).

Top left: Welsh Guards. Bottom left: Canadian Engineers. Top right: Royal Engineers. Bottom right: Royal Army Medical Corps. (Collection of Brana Radovic)

Top left: Seaforth Highlanders (51st Highland Division). Bottom left: Royal Fusiliers. Top right: possibly Sherwood Foresters or Australian 2nd Division. Bottom right: Leicestershire Regiment.

Examples of painted insignia (side). Top left: Seaforth Highlanders 51st Division. The rest are presently unknown.

Various examples of tactical signs. Top left: Unknown. Bottom left: Unknown. Top right: 18th Hussars "Queen Mary's Own" belonging to Lt. G.M. K. Young. Bottom right: Unknown.

Metal badges. Top left: Machine Gun Corps. Bottom left: Royal Horse Guards (this helmet was worn by H.R.H. King George V on his visit to the battlefront). Top right: General Officers' cap badge (this helmet belonged to General Godfrey, commanding officer of Australian forces at Gallipoli). Bottom right: West Yorkshire Regiment. This helmet belonged to Private W.T. Short V.C.

Army Chaplain's helmet with Chaplains' badge and flash of the Worcestershire Regiment, belonging to Reverend E.V. Tanner M.C. (Collection of the Imperial War Museum, London)

Helmet with metal badge and flash of the Worcestershire Regiment. (Collection of Ellwood von Seibold)

Sealed Pattern Felt Head Protection helmet for tank crews. Made by Vero & Everett Ltd., Atherstone 1916. This design was intended to replace the leather crash helmet, but seems never to have been issued. (Collection of the Imperial War Museum, London)

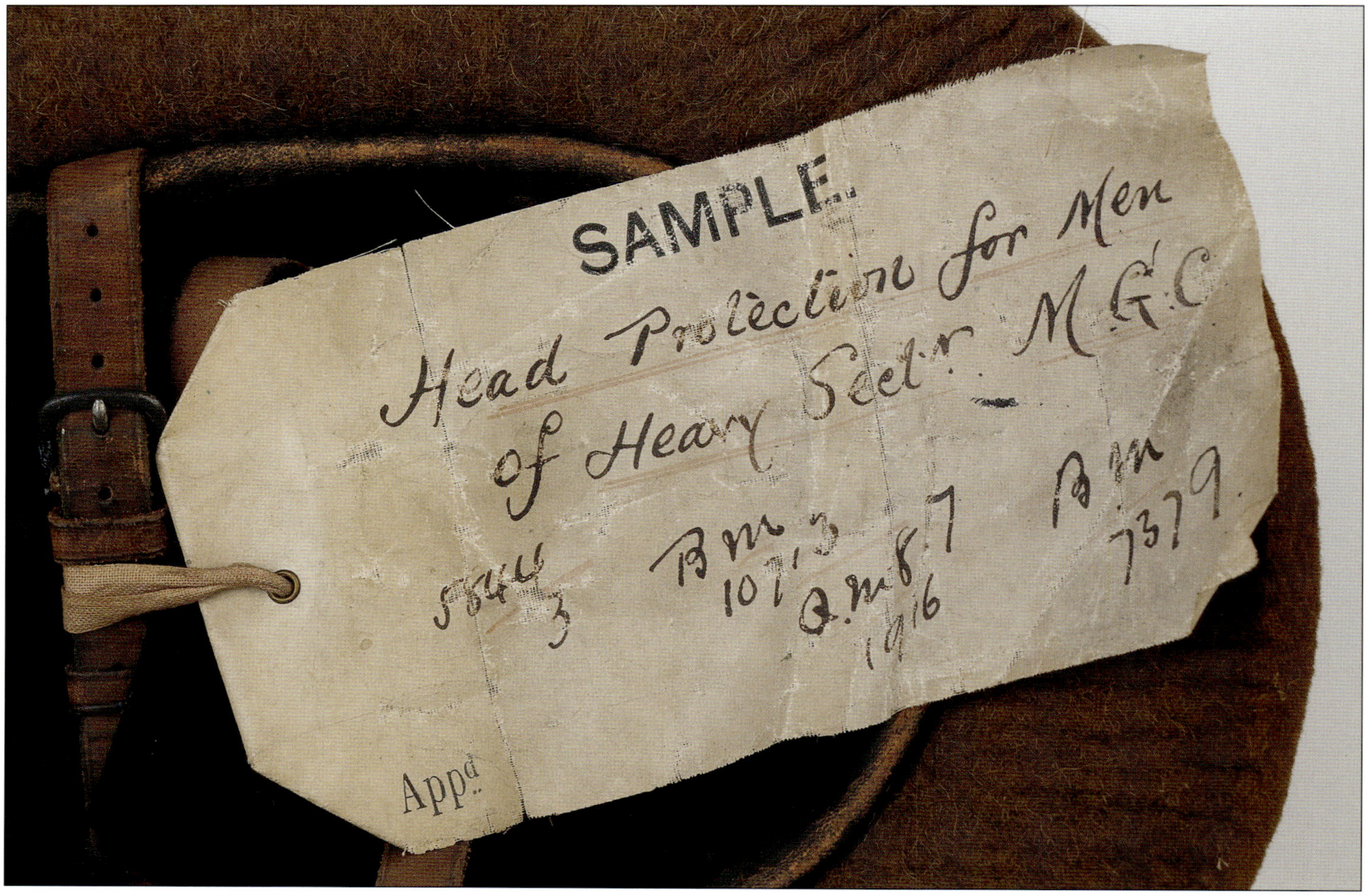
SAMPLE
Head Protection for Men
of Heavy Sectn M.G.C
Appd.

VERO & EVERITT Ltd
7
ATHERSTONE

SAMPLE
Head Protection for Men
Appd
1916

Tank Crew Protective Helmet. Leather crash helmet for tank crews. (Collection of the Imperial War Museum, London)

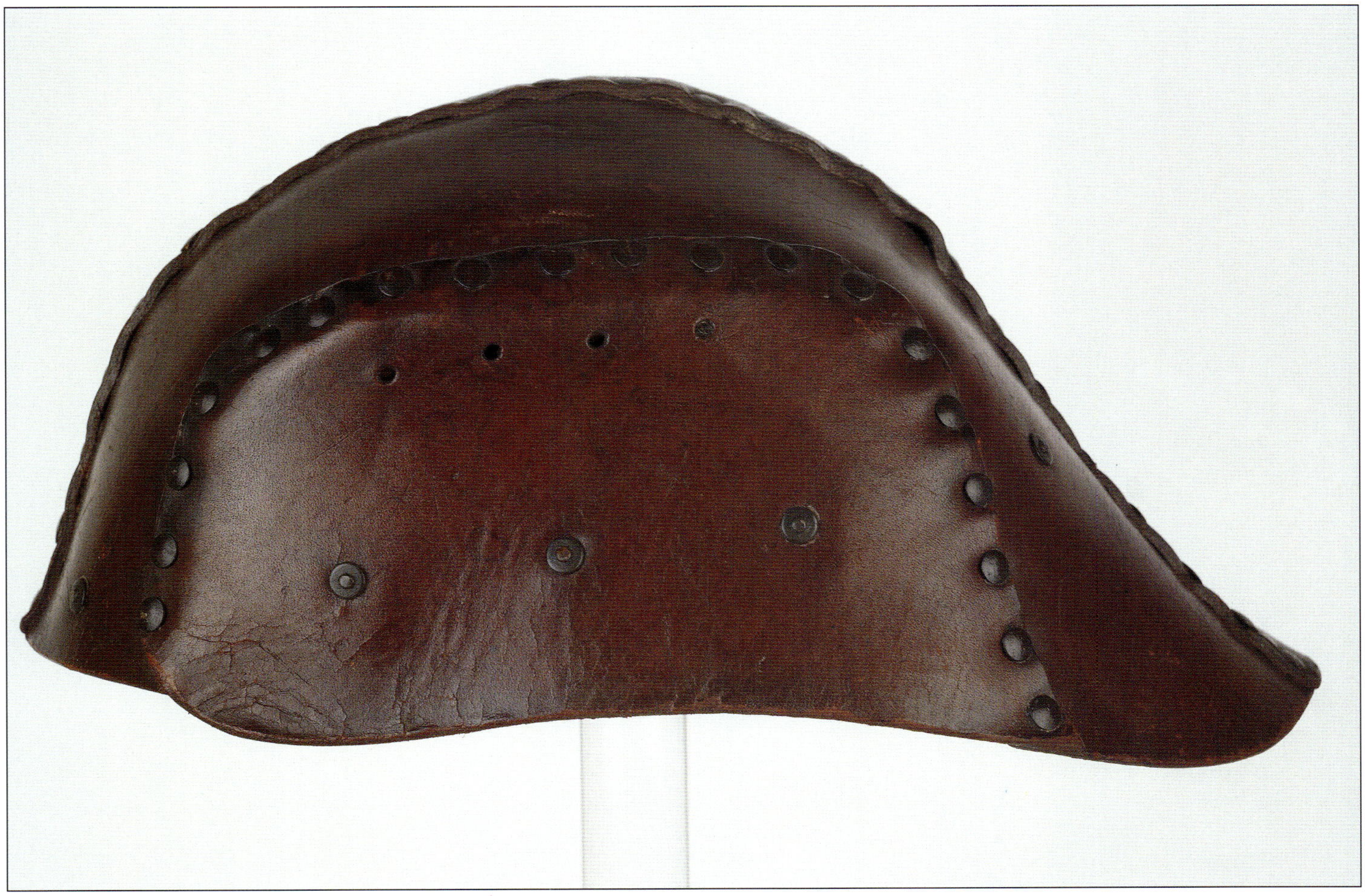

Another example of the Tank Crew Protective Helmet. (Collection of the Imperial War Museum, London)

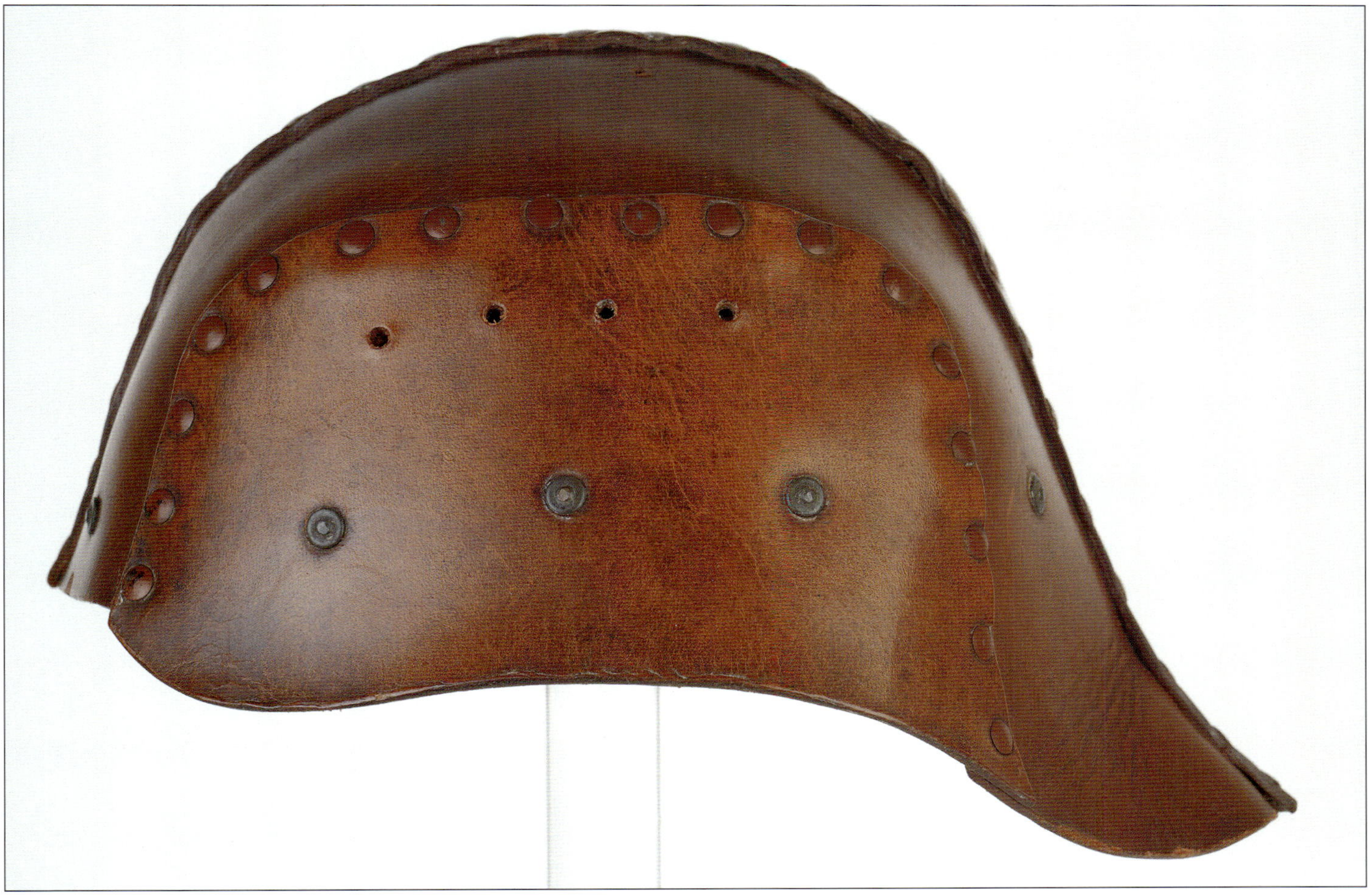

Examples of the Tank Crew Face Mask. Top: front. Bottom: back. (Collection of the Imperial War Museum, London)

■ Top: A close-up of the Tank Crew Face Mask (Collection of the Imperial War Museum, London). Bottom: Mask designed to protect against splinters and debris. (Collection of Marcus Cotton)

Examples of battle-damaged helmets.

This page and following, souvenir-painted helmets.

THE WILTSHIRE REGIMENT

THE WILTSHIRE REGIMENT
SYRIA

Batt D.O.W. WEST. RIDING.

WALTON 2/5 Batt D.O.W. WEST. RIDING.
AIR SERVICE
SOUVENIER OF THE GREAT

EGYPT
1916-19

A helmet with a metal Artillery badge together with other painted insignia, which are at present unknown. (Collection of Alan Beadle)

Note private purchase liner and post-war replacement chinstrap. (Collection of the Imperial War Museum, London)

Brodie Helmet "War Office Pattern". Note the isignia painted to the left side believed to be that of the Lancashire Fusiliers and the battle damage. (Collection of Greg Nevill)

LF

■ Top left: Paul Maze. Note the Adrian Model 1915 helmet with cloth cover. IWM Q60658

■ Top right: A British soldier wearing an Adrian Model 1915 helmet.

■ Bottom: Nurses on the French front are now issued with steel helmets. Note the Adrian Model 1915 helmets with Medical Corps badges. IWM

Top: British officer wearing a private purchase mild steel helmet with cloth cover. IWM

Bottom: British Guardsmen in a trench, Christmas 1915. Note the Brodie "Type B" helmet with camouflage paint just visible. IWM Q17391

■ Top: Sentry on a Pontoon Bridge at Robecq on which is the notice "Prepared for Demolition by Explosives". Infantry Guard 182 Bde., R.E. Party, 479 Fld-Coy. Note that he wears a Brodie "Type B" helmet with cloth cover. IWM Q11523

■ Bottom: The crater made by a Hun aerial bomb which caused many casualties. Potize. Nearest Unit 11TH BRIGADE A.I.F. 21.10.17. Note the Brodie "Type B" helmet with cloth cover in comparison with the later War Office Pattern 1915 Helmet on the right. IWM E(AUS)1096

A Brodie "Type B" helmet with four-piece cloth cover. Stephen Chambers

Two examples of the War Office Pattern 1915 Helmet. Note the nonstandard chinstrap and buckle at top.
Stephen Chambers

Stephen Chambers

Top: British prisoners of war wearing what appear to be the War Office Pattern 1915 helmet. Note the helmet liner being worn as a hat and the German guard who is wearing a pickelhaube with cloth cover. Stephen Chambers

Bottom: Decorations presented by the French to the British troops for bravery in the last Austrian offensive. Italy 24.7.18 The French Commander in Chief pinning on the Croix de Guerre with palms on a Divisional General. Note that the British officer has a metal badge attachment to his helmet and the rivets attaching the private purchase liner to the helmet shell. IWM Q26852

Top: Battle of Arras. Gordons playing cards in a reserve trench outside Tilloy-les-Mofflaines. May 1917. Note that the regimental badge of the Gordon Highlanders applied to helmet of the soldier on the right. IWM Q5225

Bottom: Officers of the 10th Bn. Royal Fusiliers, Wagonlieu, just before the Battle of Arras which began 9 April 1917. Note that the condition of the paint on the helmets would suggest that they have been newly-issued and the Chaplain's insignia. IWM Q5109

■ Top: A Staff Officer of the 36th Division outside his office dug-out at Essigny. 7 February 1918. Note the staff insignia of a lion over a crown. IWM Q10679

■ Bottom: Battle of Cambrai. 4th Battalion Gordons (51st Division) crossing a trench. Ribecourt, 20 November, 1917. Note the painted tactical signs applied to their helmets. IWM Q6278

Top: Note stencilled insignia and sliding chinstrap buckles. Greg Nevill

Bottom: Paying men of the Royal West Kent Regiment in a shell hole; near Mouquet Farm, October 1916. Note tactical flashes on cloth covers. IWM Q1429

■ Top: Battle of Arras. Royal Fusiliers (City of London Regiment). Trench Orderly room, near Guemappe. 29 April 1917. Note cloth flashes sewn onto cloth covers. IWM Q1429

■ Bottom: Moving a bath from a German dug-out. Ovillers, September 1916. Note cloth patches sewn onto the rear of cloth helmet covers. IWM Q4125

Greg Nevill

Note the metal cap badge attached to the officer's cloth helmet cover. Stephen Chambers

Note the regimental badge (Staffordshires) stencilled onto the cloth helmet cover. Greg Nevill

Note metal regimental badge (Royal Artillery) fixed to the cloth cover, and chinstrap pulled over the top of the helmet. Stephen Chambers

Top: Note the officer's Machine Gun Corps metal cap badge fixed to the cloth cover. Greg Nevill

Bottom: Brigadier-General F.W. Ramsay, commanding the Irish Brigade, 16th Division and some of his Staff. 11 June 1917. Note the private purchase cloth cover with his general's insignia attached. IWM Q6159

Note the officers' private purchase five-piece cloth covers worn by the two officers to the left. Stephen Chambers

Top: Maple Leaf Concert Party in France presenting "Alladin France". Note the cloth cover with drawstring threaded around the base of the helmet bowl. IWM CO2437

Bottom left: Examples of cloth covers. Greg Nevill/Stephen Chambers

Top: Examples of cloth covers. Stephen Chambers

Bottom: Australian troops at tea in the first support trench on the eve of battle. Westhoek Ridge. Nearest Unit – 22ND BATTALION 19.9.17. IWM E(AUS)739

Top: Sandbag defences on the coast, held by the 2nd Battalion Black Watch from the action of Arsuf, 8th June 1918, till the end of the month. Long Khaki drill trousers were worn with a short puttee round the ankle instead of the kilt during the summer to guard against risk of infection from mosquito bites. Note the four-piece tropical drawstring covers specially introduced for troops in Salonika. IWM Q12485

Bottom: French girl selling chocolate to British soldiers in the ruins of Vermelles, 27 June 1917. Note the soldier in the middle wears a "defrocked" 1st Version Cruise, with a Royal Artillery badge to the front. IWM Q5534

■ Top: HRH King George V watches the training of the Brigade of Guards before the Third Battle of Ypres (6 July 1917). The three soldiers wear the 1st Version Cruise protective helmet. IWM Q5598

■ Bottom left: Transport passing a working party on a road newly made over a mine crater on the Bapaume road. (Australians). About 19.3.1917. Note the 1st Version Cruise helmet worn by the soldier at centre right. IWM E(AUS)397

■ Bottom right: Attacking Gothas from the ground. A Lewis gun in action (Hooge). Nearest Unit 1ST ANZAC. 22.9.17. Note the 1st Version Cruise helmet worn by the gunner. IWM E(AUS)769

Top: British officers outside their dugout on the hills above Miraumont. Battle of the Ancre. IWM Q85247

Bottom: German Offensive. Collecting wounded men's steel helmets, rifles and equipment as they reach a field dressing station. Near Bapaume. 23 March 1918. IWM Q8650

Leather tank crew helmet. Andy Read

GREECE

CHAPTER NINE

GREECE

When the war started, Greece remained neutral. There was little enthusiasm for war and, in any event, sympathies within Greece were divided. On the one hand, the Greek Prime Minister, Venizelos, was in favour of the Allies. On the other hand her ruler, King Constantine I, was drawn to the Central Powers. He was, after all, married to Sophie who was the Kaiser's sister.

The Allies were hopeful that the invasion of Turkish territory at Gallipoli would bring Greece, and other countries, into the war on their side. The Greeks had long held designs on the Turkish capital, Constantinople. However, the British had in 1908 secretly promised Constantinople to the Russians in the event of victory over the Turks. Instead, the Greeks were promised the western Anatolian province of Smyrna. In the circumstances, despite the prospect of territorial gains, the Greeks were not prepared to commit troops.

However, on the 12th June 1917 King Constantine abdicated in favour of his second son, Alexandre. He, as King Alexandre I, was better disposed towards the Allies. On the 2nd July 1917 Greece declared war on the Central Powers.

The Greek army was not equipped with a suitable protective helmet when war was declared. They did not produce a helmet of their own during the war. Instead, initially the British and French respectively supplied their standard helmets to the Greeks although it appears that the majority of the helmets provided were the standard Adrian Model 1915 with the French infantry badge.

The Greek Adrian Model 1915-1918

The Greeks were anxious to standardise their equipment. In the circumstances, in 1918 the Adrian helmet was formally adopted. The French supplied the Greek Adrian Model 1915-18 helmets. Save for the badge, the helmet was identical to the standard French helmet. For further detail, please refer to the section under France.

The helmet was painted light brown. The badge was stamped out of sheet metal and consisted of a shield with a raised border with the Cross of the Greek Orthodox Church in the centre and the royal crown above. The badge measured 75mm high and 36mm wide.

It is not known how many of the helmets were supplied. Certainly, the Greek Adrian Model 1915-18 helmet in original configuration is a very rare helmet indeed today.

Following the Great War, Greece was at war with Turkey until 1922. In 1924 Greece became a republic and the Greek Model 1915-1918 Adrian helmets were converted by the removal of the royal crown from the badge.

Greek Adrian Model 1915-18 Helmet. (Collection of Ellwood von Seibold)

M.Venizelos reviewing a Greek regiment before it marches out of Salonica to meet the Bulgars. Note some of the soldiers appear to have the French infantry badge on the front of their Adrian Model 1915 helmets. IWM Q32515/Q32513

Portrait of a Greek officer wearing the Adrian Model 1915-18 helmet. Greek Army Museum, Athens

HOLLAND

CHAPTER TEN

HOLLAND

Although the Dutch Army was mobilised at the end of 1914, Holland, in common with Denmark, Norway, Spain, Sweden and Switzerland, remained neutral throughout the Great War. However, she did have a significant army, numbering some 450,000 soldiers by the end of the war.

It is perhaps worth mentioning, by way of background, that following the fall of Antwerp in 1914 some 1,600 British troops were forced to retreat into Holland. They were interned for the duration of the war. Towards the end of the war, in November 1918, retreating German troops were permitted to cross neutral Dutch territory. The price to be paid was that all military equipment, including helmets, had to be surrendered. Later, the Dutch government distributed those helmets to various organisations.

In order to defend herself and maintain her neutrality, it was important that the Dutch Army was provided with modern equipment, including helmets. In this respect, a number of experiments and tests were carried out. Two helmet models were produced and tested. The first bore a striking similarity to the French Adrian Model 1915 helmet save that, in particular, the brim was wider. The second helmet tested was similar in shape to the Adrian but had no crest and was of one-piece construction. The trials showed the helmets to be unsatisfactory.

In those circumstances, the introduction of a suitable helmet was delayed. However, the War Department was anxious to equip the troops and the French agreed to supply Adrian helmets, some 10,000 being delivered in June 1916. Those helmets were painted green and had no badge or badge holes. They continued in service well into the 1920s. For more information on the Adrian helmet, please turn to the section under France.

Tests carried out by the Dutch on the Adrian helmets showed what was considered to be a lack of strength and resistance to projectiles. Experiments therefore continued to develop a better helmet, leading to the introduction of the Dutch Model 1916 helmet.

The Dutch Model 1916 Helmet

The helmet was of one-piece construction, the steel being 1mm thick. The edge of the helmet was left plain. The helmet examined measured 292mm from front to rear and 244mm wide. There were three distinctive slots cut close to the brim of the helmet, one at each side to the front and one in the centre of the rear. These slots measured 22mm long and 4mm wide and were used to attach the helmet to the backpack. Ventilation was provided by one very small hole on each side of the helmet. The helmets were painted green.

The helmet had a German Model 1916 style liner. The liner headband was quite narrow, being only 22mm wide. The headband overlapped and was joined just forward of the left hand chinstrap attachment using two rivets. The liner headband was fixed into the helmet using eight rivets. Six of these rivets were spaced evenly around the helmet shell, three to the front and three to the rear, and were identical. The heads of these six rivets had a 7mm wide shallow dome. The other two rivets, one on each side, were used to secure the chinstrap attachments as well as the liner. The heads of these two rivets were flat and only 5mm wide. However, they had a larger, flat head on the inside of the liner than the other six. This was presumably to prevent the possibility of the rivet pulling through the liner headband due to pressure on the chinstrap.

The three liner pads each had two tongues at the end, each tongue having two holes to take the drawstring. The liner pads were approximately 122mm wide and 125mm from the liner band to the end of the tongues. A single row of stitching around the entire circumference of the headband served to secure the three liner pads. Padding was provided by means of a single piece of thick felt under each liner pad. The felt pads were secured in the same way as the liner pads with the single row of stitching. Otherwise, they were not attached to the liner pads.

The chinstraps attachments were formed from a piece of metal folded around a rectangular wire loop. The chinstrap attachments were not attached to the helmet. Rather, they were held in place between the helmet shell and liner by means of the rivets securing the liner. The loops to which the chinstrap was attached were made of heavy 3mm thick wire. Measured internally, the loops were 23mm long and 12mm wide.

The chinstrap was in two sections and was made from leather 17mm wide and 2mm thick. On the outside, it had a smooth, black finish, the inside being left rough. On each side, the chinstrap was secured permanently to the chinstrap attachments by two rivets. The chinstrap was left quite short on left and had a buckle attached to it, again using two rivets. The buckle was of iron, painted black, and was of an ingenious design. The other end of the chinstrap was left plain and was passed through the buckle and around and back through the buckle again. This created a loop and any excess chinstrap was then rolled into a tight roll and was locked in place inside the loop by the buckle and chinstrap.

Care must be taken to distinguish between the Model 1916 helmet described above and the Model 1916-22. The main difference was that the Model 1916 had a plain edge to the brim, whereas the Model 1916-22 had an edge that was rolled inwards.

Dutch Model 1916 Helmet. (Collection of Marcus Cotton)

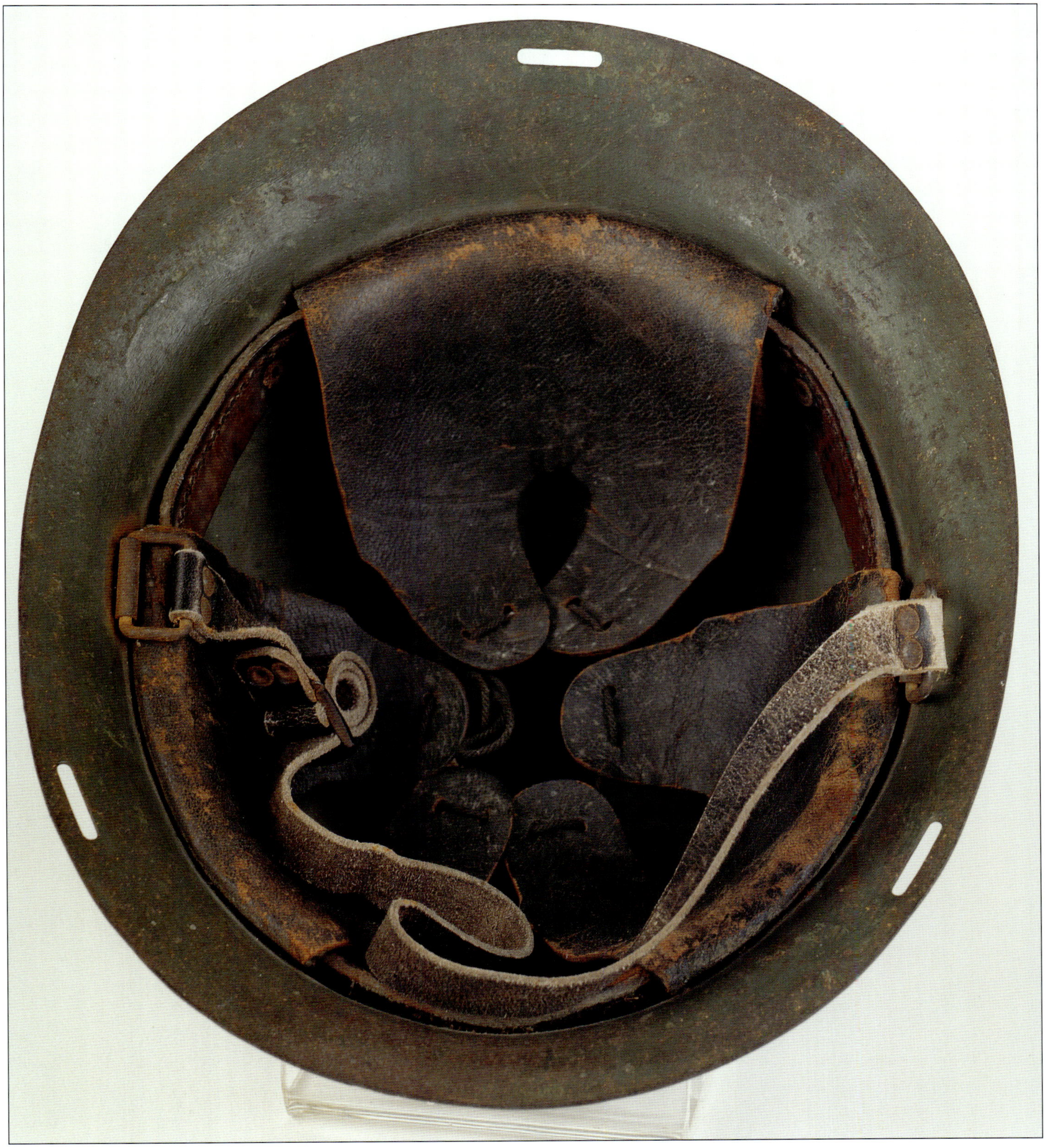

ITALY

CHAPTER ELEVEN

ITALY

Prior to the Great War Italy, in common with many other nations, was anxious to extend her territory and sphere of influence. Italy had, for instance, been victorious over Turkey in 1912 and had annexed Libya, previously a Turkish province.

However, when the war started Italy remained neutral. Although there was pressure to participate in the war, it was unclear where her best interests lay. Italy was one of the countries the Allies hoped would be drawn into the war by a successful campaign in the Dardanelles. She was promised the southern Anatolian province of Adana if she joined in the war on the side of the Allies. This offer was not sufficiently attractive and, in April 1915, Italy offered to join the Central Powers in return for substantial territorial concessions. That offer was rejected and Italy turned back to the Allies to negotiate more attractive terms for entry into the war on their side.

Terms were agreed whereby Italy would acquire substantial territory from the defeated Austro-Hungarian Empire and Turkey. The Treaty of London was concluded between the Kingdom of Italy and the Allies on the 26th April 1915. On the 23rd May 1915, Italy declared war on the Austro-Hungarian Empire and an attack was immediately launched. War was subsequently declared against the other Central Powers.

The main battles initially were waged in the South Tyrol and along the Isonzo River. There were some Italian advances but, by the end of May 1915, Austro-Hungarian resistance stiffened. As elsewhere, warfare became static, a front developing that stretched from Switzerland to the Adriatic. Much of that front crossed difficult, mountainous terrain. Fighting took place amongst rocky mountain peaks and in thick snow and ice.

A detailed history of the fighting is well beyond the scope of this book. It is worth mentioning that, on the 11th March 1916, the Italians launched their fifth battle on the Isonzo Front to help to relieve pressure on the French at Verdun. However, by November 1917 Italian reverses were such that the Italian Prime Minister, Vittorio Orlando, requested help from the French and British. The Austro-Hungarian offensive at Caporetto had pushed the Italians back in some places sixty miles to the line of the Piave River. British and French troops were sent to assist, later helping to resist the Austro-Hungarian offensives on the Asiago Plateau and Monte Grappa and on the Piave River.

It is noteworthy that American troops were not only engaged on the Western Front. On the 27th June 1918 an American infantry regiment landed at Genoa and, by the end of September, had been sent into the line. Americans were also serving in the Red Cross in Italy. In June 1918 one of their number, Lieutenant Mckey, was the first American to be killed on the Italian Front. Shortly after Ernest Hemingway, also serving in the American Red Cross, was wounded becoming the second casualty.

On the 27th October 1918, British and Italian troops crossed the Piave River, pushing the Austro-Hungarian Army back and capturing many prisoners. On the 28th October 1918 the Austro-Hungarians requested an armistice. Two days later, Hungary ended links with Austria and opened her own negotiations for an armistice. Fighting ceased on the Italian Front at 3.00pm on the 4th November 1918. Some idea of the scale of fighting on that Front may be gained by the fact that the Italians suffered some 615,000 battlefield dead.

In common with other nations involved in the conflict, Italy lacked any generally available helmet when she entered the war. Cavalry units had metal helmets for which covers were issued. However, these helmets, due to their shape and construction, were of little benefit in the conditions that quickly developed. The large number of head wounds suffered by the troops made it imperative for the Italians to obtain large numbers of helmets quickly.

The Farina Helmets

Manufactured by the engineering company of Ing. F. Farina - Via Ruffini, 10 - Milano, the Farina helmets were the first helmets issued to the Italian Army in the Great War. The helmets, together with body armour, were introduced to equip special units whose task it was to clear a way through barbed wire and other obstacles prior to an assault. Such work was extremely dangerous and efforts were made to equip the troops appropriately.

The Farina helmets came in two basic models known as the tall model and the reduced, or trench, model. The difference between the two models lies in the depth of the frontal visor. The helmets were formed from three basic components. The crown of the helmet was a shallow bowl of 1mm thick steel. To the rear was fitted a strip of steel of similar thickness. To the front of the helmet was fitted a heavy visor which was made up of four separate sheets of nickel-chrome steel riveted together using fourteen rivets. Of those fourteen rivets, four served to secure the visor to the strip at the rear of the helmet and five secured it to the front of the helmet bowl. The visor was approximately 6mm thick. The centre of the visor in the tall model was cut out at the base to allow for forward vision.

Both models exist in two different forms. In one the rear strip and visor were fitted to the inside of the helmet bowl, thus providing for ventilation. Alternatively, the rear strip and visor were fitted to the exterior of the helmet bowl, in which case there is no provision for ventilation. There was a model that was the same as the basic model save that it had an Adrian-type crest fitted to the top of the helmet bowl.

The chinstrap was of heavy leather with a buckle. It was attached on either side of the helmet just to the rear of the visor and by means of a further rivet.

The Farina helmets came in three sizes weighing from 1,680 grams for the small size trench model to 2,850 grams for the large size of the tall model. The helmets sometimes bear the Farina factory stamp, in ink, to the inside of the visor. Paint finishes varied from grey-green to dark green. In many cases the paint finish has not adhered well to the nickel-chrome steel of the visor.

The helmets were not fitted with a permanently attached liner. Rather, the helmet was worn over a greyish cotton padded cap that was separate from the helmet. In practice, often the standard issue field cap, worn reversed, or other padding was used. The helmet could be worn reversed to provide greater protection to the back of the head should circumstances warrant it.

Whereas the visor was said to be proof against the standard 6.5mm Italian service cartridge, experience quickly showed that the helmets were not proof against the more powerful German and Austrian cartridges. In this respect it is perhaps worth mentioning that in the classic Italian novel of the Great War "Sardinian Brigade" by Emilio Lusso are found several references to the unsuccessful use of the Farina helmets and body armour. This, coupled with other factors including the weight of the helmet, caused the use of the helmets ultimately to be abandoned.

The French Adrian Model 1915

The poor press received in respect of the Farina helmets caused the Italian Supreme Command to order helmets from France. The first delivery of French Adrian Model 1915 helmets was made in October or November 1915. The first helmets delivered had standard French infantry badges fitted. Subsequently, helmets were manufactured specifically for Italy and were supplied without insignia or holes for fixing insignia. Initially, the helmets were painted in the same colour as French helmets. However, later they were supplied painted in a grey-green colour.

Following the Battle of Caporetto in October 1917, the French supplied the Italians with helmets that had been manufactured as part of the order placed by pre-Revolutionary Russia. Italian troops were obliged to use the helmets with Italian insignia painted over the Russian Imperial insignia.

The Italian Model 1915

From the early months of 1916 the Italians produced helmets of their own design. The helmet strongly resembled the French Adrian Model 1915 helmet. However, the Italian helmet was manufactured in two pieces only, namely the helmet shell and crest. The crest was somewhat more pronounced than that on the Adrian helmet and was fixed to the helmet shell by six spot welds, three at the front and three at the rear, instead of rivets. There was a slot at the apex of the helmet to allow for ventilation. Approximately 5mm of the bottom edge of the helmet was folded upwards and flush with the helmet brim to provide a smooth finish.

Despite the fact that the helmet did not appear until 1916, it was designated the Model 1915. The Italian helmet had advantages in that it was easier and cheaper to manufacture and more robust than the French helmet.

The liners were manufactured from leather, leather-cloth and leather, or completely from leather-cloth with, normally, a felt headband to which the liner was attached by stitching. The liners were very similar to the French design, generally being of one-piece construction and having seven tongues. At the end of each tongue was a hole reinforced with a metal eyelet through which a drawstring passed to secure the liner at the top and provide for adjustment.

The liner was attached to the helmet using an identical method to that used on the Adrian helmet. The helmets were fitted with corrugated metal strips to provide for ventilation and size adjustment in the same way as the French helmet.

The chinstraps were made in brown leather with a single iron buckle for adjustment. One end of the chinstrap was secured to the buckle and the other to the helmet using, in each case, a single eyelet closed at one end. The chinstrap is almost identical to that on the French helmet although one example has been seen where the leather of the chinstrap is rough on both sides and another where the chinstrap is made from much heavier leather than normal. The chinstrap attachments again are very similar to the Adrian design.

The standard paint finish was a dull grey-green. The helmets appear to have been dipped in paint rather than hand- or spray-painted. Later in the war helmet covers were issued in various colours and types of material.

The Italians made limited use of Italian Model 1915 helmets fitted with Dunand visors. Additionally, they experimented with metal shields, suspended by a strap around the bowl of the helmet, and hanging down on each side of the face to give added protection. These were known as Lippman cheek-guards. Their use appears to have been very limited. Some use was also made of the French brow-plate and other non-official additional strengthening to the helmet.

It was a common practice to paint unit identification badges and/or numbers to the front of the helmet. Rank insignia was also worn to the front or side of the helmet. During the Great War there do not appear to have been any official regulations in respect of helmet insignia. However, the practice of adorning helmets and helmet covers with painted metal or cloth insignia appears to have been widespread. For instance, the infantry would paint the number of their regiment usually, but not always, in black with the royal crown above.

Austro-Hungarian Use of Captured Italian Helmets

The Austro-Hungarians made use of captured Italians helmets. For further details, see under Austro-Hungary.

Farina Helmet: Reduced or Trench Model. Note insert showing manufacturers' mark. (Collection of Marcus Cotton)

Farina Helmet: Tall model with chinstrap missing. (Collection of Michael Haselgrove)

French Adrian Model 1915 Helmet as supplied to the Italians. Note absence of badge and badge attachment holes. (Collection of Ellwood von Seibold)

French Adrian Model 1915 Helmet as supplied to the Italians. Green painted and with the 225th Infantry Regiment badge painted to the front. (Collection of Brana Radovic)

■ Above and top right: Standard Italian-manufactured Model 1915 Helmet. (Collection of Michael Haselgrove)
Bottom right: For comparison purposes; a French Adrian Model 1915 Helmet. (Marcus Cotton)

Comparisons of the Italian Model 1915 Helmet.

Italian Model 1915 Helmet. Note green paint finish, painted unit insignia to front and lieutenants' rank to the left-hand side. (Collection of Marcus Cotton)

Various Italian Model 1915 Helmet liners. Top left: One-piece leather cloth liner. Note ink stamp consisting of a crown surmounting the initials MA within a circle, believed to be a manufacturers' mark. Note also that the chinstrap of brown leather, has been left rough and unfinished on both sides. Bottom left: Two-piece liner with leather headband and leathercloth tongues. Note unusually heavy gauge leather chinstrap. (Collection of Michael Haselgrove) Right: Seven-piece leather liner consisting of headband and six separate tongues. Note size ink stamp on liner and British government acceptance label. (Collection of Imperial War Museum, London)

Six-piece blue cloth officers' cover with embroidered Alpini insignia. Note the 2nd lieutenants' rank insignia. (Collection of Brana Radovic)

Various covers. (Collection of Brana Radovic)

Leather carrying harness. (Collection of Brana Radovic)

Damaged Italian Model 1915 Helmet. The damage, it is thought, was inflicted by the British for test purposes. (Collection of Imperial War Museum, London)

Top: Italian engineers wearing the (deep) Farina armoured helmet and body armour.

Bottom: Two officers of the Sardinian Brigade (Sassara) conversing. Troops manning position on a road, 1918. Note painted rank insignia of the officer on the left and the stencilled unit insignia on the helmet of the other officer. IWM Q65258

Italian soldier wearing the French-supplied Adrian Model 1915 helmet with an unknown insignia stencilled to the front.

Top: Italian troops of the 35 Infantry Division marching through Salonica, August, 1916. Note the French-supplied Adrian Model 1915 helmets with stencilled insignia of the 63rd Infantry Regiment. IWM Q32261

Bottom: New type of Italian Machine Gun. (Pistola Mitragliatrice). Note the Italian Model 1915 helmets with what appears to be the stencilled badge of the 30th Infantry Regiment. IWM 56999

An Italian motor cyclist of the Bersaglieri with an Indian machine. Note that he wears the traditional feathers of the Bersaglieri attached to his Italian Model 1915 helmet. IWM Q26037

Top: Italian "Arditi" or shock troops who took a leading part in the present offensive. Note the stencilled insignia. The soldier on the right wears his helmet reversed. IWM Q54781

Bottom: Battalion Headquarters at Casa Ricci, 1918. (Probably near Candelu, Piave Front). Note the cloth badge stitched to the cloth cover of the soldier on the left. IWM Q65346

JAPAN

CHAPTER TWELVE

JAPAN

Prior to 1914, Japan had concluded a treaty with Great Britain and early in August 1914 reiterated that she was prepared to do her part to aid Great Britain, as was laid down in the Anglo-Japanese Alliance.

On the 15th August 1914 the Japanese presented an ultimatum to Germany requiring the surrender of German territory on the Chinese mainland. The deadline for a reply was the 23rd August 1914 and, when no satisfactory reply was received, Japan declared war against Germany. A Japanese force promptly set out to capture the German Protectorate of Kiaochau on the Shantung Peninsula on the Chinese mainland where the Germans had constructed their fortified naval and military stronghold of Tsing-Tao. The landings took place on the 2nd September 1914. British and Indian forces were also involved in the operation, Tsing-Tao being formally occupied on the 16th November 1914.

With the declaration of war, the Pacific Ocean island groups acquired by Germany prior to the war were vulnerable. By the end of September 1914, British, Australian and Japanese troops were moving to occupy them. On the 3rd December 1914, the Japanese Government obtained a promise from the British that Japan could occupy all German territory north of the Equator. This angered the Australians who had aspirations in respect of the territories involved.

When the war ended, all German territory in the Far East had been conquered. None of these territories were returned to Germany. Instead, it was decided to establish a system of League of Nations Mandates. Mandates were given to victorious powers, as was done in the Middle East. Japan acquired the Mandate for the Marianas, Caroline and Marshall Islands. The Japanese also, against the wishes of the Allies, annexed the Chinese province of Shantung.

Prior to, and in the early part of the Great War, Japan was far better prepared than most of the Allies. She was able to provide military stores for many of the Allied nations. During the course of the war, huge stocks of armaments were provided by Japan, particularly to Russia. Something like 750,000 rifles, as well as artillery, ammunition and clothing were supplied to Russia alone. Britain too received considerable assistance, rifles in particular being supplied for use by the expanding armed forces, including the Navy.

The Japanese Navy assisted in the campaign in the Pacific against German cruisers and helped in the escort of Australian troops to Europe. Japan provided a destroyer flotilla to operate against German submarines in the Mediterranean and Japanese transport ships were provided to the United States to assist in the transport of war materiel and troops to Europe.

Although the Japanese had carried out considerable experimentation with trench shields and other types of defences prior to the Great War, they did not have any helmet in general service at the start of the war. The introduction of the French Adrian Model 1915 appears to have prompted Japanese experimentation and, in 1917, a helmet was introduced.

The Japanese Model 1917 Helmet

The helmet bore a strong resemblance to the Adrian and Sohlberg helmets. However, the Japanese helmet had a more pronounced visor and a much smaller neck-guard. The reduced neck-guard may or may not be regarded as an improvement on the other helmets. There were complaints that the neck-guard on the Adrian interfered with the backpack, particularly when shooting in the prone position. However, the reduced neck-guard clearly provided less protection to the wearer.

There was a ventilation hole at the apex of the helmet. This was covered by a concave piece of metal in the form of a stylised cherry tree flower with five petals. The badge on the front was a five-pointed metal star.

The helmet appears to have been first used in June 1917. Further details of this helmet are unknown and the Authors have been unable to examine an example.

The Japanese Model 1918 Helmet

Following the introduction of the Model 1917 helmet, the Arsenals at Tokyo and Osaka were asked to develop a new helmet. The new helmet was very similar in external appearance to the earlier helmet and was designated the Model 1918.

The following remarks and measurements are based on what is believed to be a high-quality reproduction Japanese Model 1918 helmet.

The first design of the Model 1918 had fifteen small ventilation holes in the apex of the helmet. These were in the form of a star and were not covered. On the second model, the single, circular ventilation hole was covered with a metal plate. As with the Model 1917 helmet, the covering plate was in the form of a concave cherry tree flower with five petals. It measured 66mm across and was secured with five small, dome-headed rivets.

The helmet examined measured 283mm from front to rear and 235mm from side to side. The naval badge measured 30mm high and 23mm wide. Approximately 5mm of the helmet rim was folded inwards and pressed flat to provide a smooth finish.

The liner was very similar indeed to the German Model 1916 helmet. The Japanese helmet had a leather headband. This was 43mm deep and was secured at the rear by stitching. The headband was secured to the helmet by means of three split pins, one on each side at the front and one at the rear centre. The split pins had slightly domed heads measuring 10mm across. Apart from the slightly larger domed heads on the Japanese helmet, the split pins were almost identical to those used on the German helmet.

The liner had three liner pads in the same form as the German liner, although the Japanese pads were made from two separate pieces of leather stitched together. Each liner pad ended with a double tongue with a hole in each tongue to take the drawstring. The liner pads measured approximately 140mm wide. On the inside of each liner pad was a pocket constructed from white linen material and into which padding was placed. Again, these were very similar to the German liner save that on the Japanese helmet each liner pocket was secured with two separate sets of cotton tape, each tied with a bow.

The chinstrap was very different to that used on other helmets. It was formed from two separate pieces of woven (light webbing) tape 18mm wide. Each tape was attached at two points to the side interior of the helmet. These chinstrap attachment points were quite widely spaced, the distance between the outside of each pair being about 146mm. Each chinstrap attachment consisted of a piece of rectangular-shaped metal folded around a wire loop. The wire loops were 22mm wide and were very similar to those used on the Adrian Model 1915 helmet. Each chinstrap attachment was fixed to the helmet by means of two rivets of the same type used to secure the ventilation hole cover. The gap between the pairs of rivets securing the chinstrap attachment to the helmet was 6mm.

To attach the chinstrap tapes, one end was first passed through the two attachment points on one side of the helmet. It was then stitched to the tape, forming a triangle and attaching the chinstrap permanently to the helmet. That left one end of the tape, measuring approximately 460mm long, free. There was no buckle, the ends being secured by tying under the chin.

The helmets were painted a yellowish-brown for the Army and grey-blue for the Navy. The Army helmet was fitted with a badge in the form of a five-pointed star. The Navy and Marine helmet had a badge in the form of an anchor with a cherry tree flower at the centre.

Although the Model 1918 helmet was developed during 1918, it is not known when the helmets were first introduced into service. It is thought that the helmets were not used in actual combat until 1927.

■ Japanese Model 1918 Helmet with Marine insignia. The owner points out that this is probably an accurate, high-quality reproduction. (Collection of Marcus Cotton)

POLAND

CHAPTER THIRTEEN

POLAND

At the start of the Great War Poland was divided between three dominant powers, namely Germany, Austro-Hungary and Russia. However, there was a strong Polish national sentiment. The Polish people had never given up their dream of an independent Polish state. To some extent, Poles fought on the side controlling the area where they lived. However, where there was a choice they fought for the side most likely to grant their wish for their own state.

Various individuals established paramilitary organisations. These lead to the creation of Polish units, or legions, in the Austrian, German and Russian armies. Later in the war, a Polish Legion was formed within the French army. The major powers sought to attract Polish nationalistic fervour and the different Polish Legions to their cause. For instance, the Russians early in the war published two manifestos. One promised a re-born free Poland after the war. The other called upon the subject peoples of the Austro-Hungarian Empire to rise up and demand their independence. The Russian Legion, led by Pulawy, was established.

On the other hand, the Austro-Hungarians encouraged the Polish Legion, led by Joseph Pilsudski, to regard that Legion as the foundation of a Polish National Army. Later, the Germans set up a Polish military force used largely to garrison territory taken from the Russians.

On the 5th November 1916, the Germans proclaimed the foundation of the Kingdom of Poland. That Kingdom was to have Warsaw as its capital and was to have an army with German officers in command. However, no political power or government came with the creation of the country. Joseph Pilsudski refused to co-operate. The German plan, never sincere in its intentions, was unsuccessful.

On the 21st January 1917 President Wilson of the United States appealed for a united Poland to emerge from the war as a sovereign state. The Tsar of Russia supported that scheme, but whether his support was sincere is questionable.

The Allies had captured considerable numbers of Polish troops fighting on the side of the Central Powers. They were eager to fight for the liberation of their country, their numbers being increased by Polish volunteers who came, in the main, from the United States and Canada. The French were short of men and that shortage was made worse by mutiny in the French Army. On the 4th June 1917 a decree was passed creating a Polish Army within the French Army and under French command.

In July 1917, Polish troops serving in several Polish Legions within the German army, refused to take the oath of loyalty to the Kaiser. Joseph Pilsudski and a number of his troops were arrested and interned for the remainder of the war.

On the 8th January 1918 President Wilson, in his address to the United States Congress, included in his fourteen point plan that a Polish State should be created. It would be united, independent and autonomous with free, unrestricted access to the sea.

On the 3rd March 1918 the Treaty of Brest-Litovsk was signed between Russia and Germany. The Bolsheviks gave up all their interest in Poland. However, despite this success in the east, the tide was turning against the Central Powers. On the 3rd June 1918 Britain, France and Italy announced their full support for Polish statehood. On the 7th October 1918, the Regency Council in Warsaw declared a free, independent Polish State. The declaration was challenged. The Ukranians did not wish to lose East Galicia to the new Polish State and fighting erupted. However, the new State had been born.

On the 10th September 1919 Austria signed the Treaty of St. Germain. Poland was given the former Austro-Hungarian provinces of Western and Eastern Galicia, including Cracow and Lemberg.

In the summer of 1920, a brief war was fought between Soviet Russia, the Polish and the Ukrainians. All three armies used, to a greater or lesser extent, the Adrian helmet. However, numerous alterations were made to the helmets, both during and after the conflict. The alterations included the removal of the original badge and the substitution of a new badge, either painted or in metal. Some helmets were repainted, as mentioned elsewhere. Poland continued to use the Adrian helmet even after the introduction of their new helmet years later. The fact of the fighting mentioned above, and other conflicts, coupled with the continuing use of the Adrian helmet by the Polish Army, helps to explain why the helmets, in original configuration, are so scarce today.

German and Austro-Hungarian Polish Legions

The Polish Legion, largely recruited from Galicia, fighting with the Austro-Hungarian Army wore standard German or Austrian helmets, as worn by Austro-Hungarian troops. The helmets did not have any distinctive badge or other feature and it appears that the practice of painting a white eagle, or fixing a metal eagle, to the front of the helmet developed after the end of the Great War.

Similarly, Polish Legion troops fighting with the German Army wore the same helmet as the German troops.

Russian Polish Legion Helmets

Poles fighting in the Russian Army were equipped with standard Russian equipment, including helmets. However, following a decision reached in March 1918, the 3rd Infantry Division was issued with modified Russian Model 1915-16 Adrian helmets. The standard Russian badge was removed and replaced with a white painted or metal eagle. In addition, a one centimetre high band of dark red paint, representing the Polish national colour, was painted around the base of the helmet bowl. Two other infantry divisions, reserve troops and foot artillery later adopted the helmet in this configuration. The coloured band was sometimes repainted the colour of the particular unit. For instance, the foot artillery adopted a black band within a red border. At the bottom of the eagle badge there was sometimes a small shield where a unit symbol could be inserted. On officers' helmets, the eagle was sometimes painted in silver.

French Polish Legion Helmets

The Polish Legion was issued with standard French Adrian Model 1915 helmets. Mostly, these had one of two badges. The first was the French Chasseurs' badge with a Polish eagle stamped in the centre of the horn. The second was the Polish eagle stamped on an oval plaque. Most of the helmets issued appear to have the Chasseurs' badge. In addition, there was limited use of the French Adrian Model 1915 helmet with the artillery badge. For further information, see the section under France.

Polish units operating in Northern Russia were equipped with standard British helmets. These helmets sometimes had a painted eagle to the front.

French Polish Legion Chasseurs' Adrian Model 1915-17 helmet. (Collection of Ellwood von Seibold)

Various examples of the French Polish Legion Chasseurs' Adrian Model 1915-17 helmet.

French Polish Legion Adrian Model 1915-17 helmet used by other troops apart from Chasseurs. (Collection of the Polish Army Museum, Warsaw)

Adrian Model 1915-16 helmet used by the 1st Polish Corps fighting in Russia under General Musnicki. Note the silver painted eagle insignia denoting use by officers over the mark left by the removal of the original Russian badge. (Collection of the Polish Army Museum, Warsaw)

Adrian Model 1915-16 helmet used by the 1st Polish Corps fighting in Russia under General Musnicki. This example has a white painted eagle insignia denoting use by other ranks. Also note the mark left by the removal of the original Russian badge. (Collection of the Polish Army Museum, Warsaw)

3912.8 MWP

Adrian Model 1915-16 helmet. This officers' example has a number "2" painted under the eagle denoting its use by a member of the 2nd Division, 1st Polish Corps. (Collection of the Polish Army Museum, Warsaw)

Adrian Model 1915-16 helmet used by an officer of Engineers serving with the 1st Corps. (Collection of the Polish Army Museum, Warsaw)

Russian Adrian Model 1915-16 helmet. Note that the original Russian metal badge has been removed and the 1st Polish Corps other ranks' painted insignia substituted. (Collection of Brana Radovic)

British Helmet Steel Mark I used by Polish Forces in Northern Russia. (Collection of the Polish Army Museum, Warsaw)

Extemporised armoured train. Steel helmet appears Polish possibly North Russia 1919. Note the white painted Polish insignia to the front of the British Mark I helmets. IWM Q80062

PORTUGAL

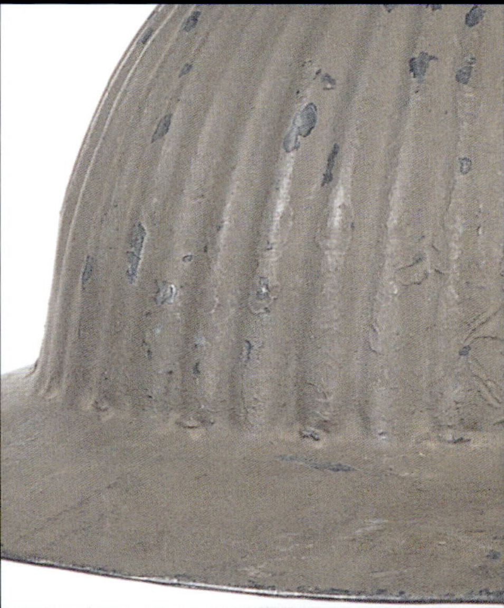

CHAPTER FOURTEEN

PORTUGAL

Before and during the Great War, Portugal was in political turmoil. She became a republic on the 5th October 1910. However, that did not bring political stability, there being no less than eleven governments between 1910 and 1916. Portugal had a long-standing association and ties with Britain, although these became less strong when Portugal became a republic.

In the circumstances, the start of the Great War did not directly threaten Portugal. The Portuguese Government did affirm her loyalty to Britain in August 1914, but no treaty or obligation bound her to enter the war and she remained neutral.

The Portuguese had overseas interests, including the colony of Portuguese Mozambique. It was inevitable that German and Portuguese interests in Africa would lead to conflict. With an attack on the Portuguese colony, and the spread of the fighting in East Africa, Portugal could not avoid becoming involved in the Great War. Ultimately, Germany declared war on Portugal on the 9th March 1916.

Although weakened, Portuguese association with Great Britain stretched back over many centuries. It was therefore to the British that Portugal turned for assistance in the form of military supplies and expertise. The British were happy to oblige.

Portuguese mobilisation was ordered, following lengthy preparation, on the 18th December 1916. The first contingent of the 40,000 strong Portuguese Expeditionary Force landed in France on the 3rd February 1917. The Portuguese troops entered the front-line for the first time in the Neuve Chapelle Sector on the Western Front on the 30th May 1917. They were in action for the first time in June 1917. Later, in April 1918, the full force of the German Offensive on the Lys River fell on the Portuguese Second Division.

The only helmet in use in August 1914 with the Portuguese armed forces was made of hardened felt and lined with cork and had been designed for use in the colonies. The question of the supply of a protective helmet for the Portuguese troops who were to fight on the Western Front became urgent and it was to the British they turned for assistance.

The Portuguese Model 1916

It has been said that the Portuguese military authorities were attracted by the design of a Portuguese helmet made at the end of the 15th century. That helmet had a ribbed bowl and the Portuguese wished to introduce a similar design. To that end, some say they approached the British for assistance in the design and manufacture of such a helmet. Whilst interesting, the authors consider the theory incorrect. The helmet introduced actually bears little resemblance to the 15th Century helmet. The Portuguese Model 1916 much more closely resembles the design of private purchase helmets used by the British earlier in the war. For more information on the private purchase British helmets, turn to the section under Great Britain.

The approximate dimensions of the Portuguese helmet were 128mm high, 308mm long and 257mm wide. The helmet brim measured approximately 42mm wide at the sides and 48mm wide at the front and rear.

On each side of the helmet bowl there was a ventilation hole approximately 80mm above the brim. A circular washer-like piece of metal, crimped on the inside of the helmet bowl reinforced each ventilation hole. Folding downwards and inwards of the helmet brim provided a smooth finish.

The liner consisted of a leather headband 49mm deep attached to a simple woollen cap-like lining fitted closely to the helmet bowl. Around the helmet between the helmet shell and the liner may be found a strip(s) of thin, corrugated cork. The liner attachment to the helmet was by means of a cord or a double thickness of twine passing around the base of the helmet bowl through a series of holes in the helmet shell. The holes were punched in pairs at the base of every other raised rib on the helmet bowl. The chinstrap was made of leather approximately 10mm wide. Badges were sometimes painted to the front of the helmets, although it appears that the practice was discontinued later in the war. Badges observed consist of crossed rifles over an oval shield on which was shown the number of the regiment. Another badge consisted of crossed sabres with the unit number shown above. Painting of the helmets was by hand in either a grey-blue or light brown paint.

The British were already aware of the defects of the Portuguese Model 1916 helmets when they were issued. In particular, the mild-steel used for the helmet shells had poor protective qualities against projectiles of even moderate velocity. The ribbed design did not, as was intended, strengthen the helmet. Rather, it tended to allow projectiles to catch the surface and penetrate instead of being deflected. It is probable that the Portuguese troops were issued with these helmets as a temporary measure.

The Portuguese Model 1917

At the end of 1917, the British proposed they provide the Portuguese with standard British Mark I helmets these, by that stage, being available in sufficient numbers. When these helmets were provided, they were known as the Portuguese Model 1917.

Once the Model 1916 helmets had been replaced, they were used to supply the Portuguese Civil Defence and other organisations. Both types of helmet continued in service for a number of years.

Portuguese Model 1916 helmets. (Left: Collection of Imperial War Museum, London. Right: Marcus Cotton)

■ Portuguese Model 1916 helmet. Top view and interior. (Collection of Imperial War Museum, London)

Wounded Portuguese artilleryman. Note the Portuguese Model 1916 helmet suspended by the chinstrap to his shoulderstrap. The corrugated cork of the liner band is just visible. IWM Q60286

■ Top: Portuguese Mule Limber with a load of shovels. Nr. Locon, June 24th, 1917. Note that the soldier on the left wears the Portuguese Model 1916 helmet, whereas the officer wears the Portuguese Model 1917 helmet. IWM Q5544

■ Bottom: Scene on the Portuguese Front. France. Entrance to a dug-out showing an anti-gas curtain. Note that Portuguese Model 1916 (right) and Model 1917 (left) helmets are shown. IWM Q64451

ROMANIA

CHAPTER FIFTEEN

ROMANIA

Prior to August 1914 Romania was aligned with the Central Powers, although sympathies within the country were divided. Culturally she had much in common with France and it was, in any event, not clear where her best interests lay. In the circumstances in August 1914, along with Italy, Portugal, Greece, Bulgaria and Turkey, Romania remained neutral.

Romania was one of those countries that Britain hoped would be brought into the war on the side of the Allies by the campaign in the Dardanelles. It was argued that a successful campaign would give the Allies the River Danube as a line of communication for an army penetrating into the heart of the Austro-Hungarian Empire. It would enable British sea power to be brought to bear in the middle of Europe. Lloyd George, the British Prime Minister, summarised the benefits of the Dardanelles campaign in a memorandum to the War Council, saying:

> *"To bring Bulgaria, Romania and Greece in with Serbia means throwing an aggregate army of 1,500,000 on to the Austrian flank. This will not only relieve the pressure on Russia, but indirectly on France. It will tend to equalise things, and thus give us time to re-equip the Russian Army"*.

The offer made to Romania, if she came in on the side of the Allies, was that she might receive from the victorious Allies a port on the Sea of Mamara. Romania was not persuaded. She remained neutral, whilst at the same time allowing German and Austro-Hungarian personnel and war materiel to pass through her territory en route to assist Turkey.

However, the position was changing. King Ferdinand I had succeeded to the Romanian throne in late 1914. He was better disposed towards the Allies than his predecessor and negotiations began. There were high hopes in 1916 for the continuing Russian offensive against Austro-Hungary. The promise was made to Romania that if she joined the Allies and success followed, she would receive part of the Austro-Hungarian provinces of Transylvannia and Bukovina and the Banat region in its entirety. Negotiations led to a secret Treaty, signed on the 18th August 1916, between Romania, France, Italy and Great Britain. France agreed to supply Romania with military equipment, including helmets.

On the 27th August 1916 Romania declared war on Austria. Romanian troops moved into Transylvannia. It was this offensive, coupled with other factors such as the Russian offensive in the east and the Anglo-French offensive on the Somme in the west, that persuaded the Kaiser that it was appropriate to allow the resumption of unrestricted submarine warfare. The resumption of unrestricted submarine warfare on the 1st February 1917, involving as it did the sinking of neutral shipping, was a major factor in bringing the United States into the war on the side of the Allies.

The Romanian advance into Transylvannia was initially successful. However, on the 15th September 1917, Hindenburg ordered that all German efforts were to be directed against Romania. A swift German advance followed and Romania suffered severe losses. Romanian forces were pushed back over the Transylvannian Alps and across the Romanian border. The Government was forced to evacuate the capital city of Bucharest.

Despite reverses, the Romanians fought on, fortified by further promises by the Allies. In particular, President Wilson addressed the United States Congress on the 8th January 1918. He set out his fourteen-point plan for a peace programme in Europe. This included the restoration of Romania and an association of nations to help guarantee political independence and territorial integrity following the war. The peoples of Austro-Hungary would get, if not independence, then a free opportunity for autonomous development.

However, following the Russian Revolution and the signing of the Treaty of Brest-Litovsk between Russia and the Central Powers, the Romanian position was irretrievably compromised. On the 5th March 1918 Romania signed a treaty with the Central Powers followed, on the 7th May 1918, by the signing of the Treaty of Bucharest. Romania was obliged to cede territory to Bulgaria although, in return, she was offered Bessarabia if she could eject the Bolsheviks.

Following the end of the war, Austria signed the Treaty of St. Germain pursuant to which Bukovina was ceded to Romania. Bulgaria signed her peace treaty at Neuilly in France. Pursuant to that Romania received the transfer back of Dobruga, a small strip of land along the Black Sea shore.

The Romanian Adrian Model 1915-17 Helmet

At the end of 1916, Romania ordered 200,000 French Adrian Model 1915 helmets. Of those, the French were able to deliver approximately 90,000. The helmets were known as the Romanian Adrian Model 1915-17. They were painted blue and were, apart from the badge, identical to the French helmet. For further information, see the section under France.

The Romanian badge was the royal cipher of King Ferdinand I. It consisted of two stylised "Fs" placed back-to-back with a central numeral 1 surmounted by the royal crown. The badge was 73mm high and 57mm wide and was unusual in that it consisted of an ovoid shield rather than having the metal trimmed to the shape of the badge. The badge was contoured precisely to the shape of the helmet and was attached by means of two prongs in the same way as other Adrian helmet badges.

It is estimated that Romania suffered some 335,000 war dead. Although the sympathies of Romanian nationals differed, and Romanians fought on the side of the Central Powers as well as for the Allies, certainly the number of helmets provided by the French was inadequate to provide every soldier with a helmet. Some improvisation by the Romanian Army was necessary. When the Russian Revolution commenced, the Romanians appropriated Russian Adrian helmets from mutinous Russian troops. One will thus occasionally come across a Russian brown-painted Adrian to which has been attached a blue Romanian badge.

It is worth commenting that in 1923 the unified Romanian army adopted khaki as their standard uniform colour. The Romanian Adrian Model 1915-17 helmets remained in service and were painted khaki to match the uniform colour. The interiors appear to have been left in the original blue colour.

In 1930 King Carol II succeeded to the throne. A new emblem was adopted. This consisted of his personal cipher, two letter "Cs" with one reversed. The conversion of existing helmets by the substitution of the new emblem as a helmet badge appears to have been very limited.

Romanian Adrian Model 1915-17 helmet. (Collection of Ellwood von Seibold)

Romanian Adrian Model 1915-17 helmet. (Collection of Marcus Cotton)

A trench occupied by Romanian troops, 1917. They appear to be wearing the Romanian Adrian 1915-17 Helmet.

■ Top: Battle of the Sereth. Prisoners captured during the battle being marched to a collection station. Romanian front, 1917. IWM Q60324

■ Bottom left: Battles of Merasesti, August 1917. A Romanian soldier wounded during the battle being brought in on an improvised stretcher. Foscani, August 1917. Note that the soldiers in this photo and the one above (which appears to have been taken at the same place) wear the Romanian Adrian Model 1915-17 Helmet. IWM Q60356

■ Bottom right: Anti-aircraft machine-gun. Hano-Konaki, 1917. IWM Q58118

Top: Romanian officers wearing Romanian Adrian Model 1915-17 helmets.

Bottom: Group of Divisional Staff Officers, Mochvunoille, 1917. IWM Q50386

RUSSIA

CHAPTER SIXTEEN

RUSSIA

A detailed account of the Russian involvement in the Great War is well beyond the scope of this book. However, by way of background, suffice it to say the position in August 1914 was that a treaty signed in 1892 linked Russia and France. Britain had entered into the Entente Cordiale with France in 1904 and had signed an agreement with Russia in 1907. The three countries, including their respective empires and dominions, were thus allied to one another.

Within Germany there existed jealousy and resentment of the success of the three counties and their empires, together with a fear of encirclement by the powerful allied nations to the east and west. Further tension was caused by the Russian support for Serbian and Slav unification at the expense of Austro-Hungary, Germany's ally.

When war came between Austro-Hungary and Serbia it was not inevitable that other nations would be involved. Diplomacy might have prevailed. However, the mobilisation of armies, and the great difficulty in stopping that mobilisation once started, coupled with a lack of effective diplomacy and the system of treaties and alliances in place, ultimately served to make war between the great powers inevitable.

When the Great War started the huge size of the Russian Empire entering the war against Germany and Austro-Hungary caused great excitement in Britain and France. The Russian Army was likened to a steam-roller. There were rumours of a vast Cossack Army on its way to Northern France via Britain.

The reality was somewhat different. Although Russia did have a large army, and there had been a great improvement in that army since the Russo-Japanese War in 1905, the problem was that the vast amount of equipment required upon mobilisation was simply not available. Even if the army had been fully equipped, the problems involved in transporting the army over the vast distances involved were almost insurmountable. Unlike the railway network within Germany, designed as it was for military purposes enabling large numbers of troops to be moved along the borders and across the country in time of war, the Russian road and rail networks were woefully inadequate or non-existent.

The Russian Army in peacetime numbered about 1,300,000 men. When that force was brought up to its war footing it numbered about 2,000,000 men, with an equal number in reserve. Huge amounts of financial and material assistance had to be provided, particularly by Japan, France, Britain and the United States. It soon became clear that the war against the Central Powers would require a joint effort from the Allies and reliance could not be placed on any one country, even one as large as Russia.

Apart from the huge involvement of the Russian Army on the Eastern Front, Russian troops fought on other fronts. For instance, on the 20th April 1916 Russian troops landed at Marseilles to support the Allies on the Western Front. The French received them with rapturous enthusiasm. Initially, they fought alongside the French in the Champagne Sector. In April 1917 they were in a successful attack at Courcy and on the Aisne-Marne Canal. On the 30th July 1916 Russian troops arrived to fight with the Allies on the Salonica Front. They were destined to form part of the Army that was to liberate Serbia.

When the Great War commenced, there was no generally available Russian helmet. Russian troops were equipped with cloth caps or fur hats and, judging from the photographic evidence available, that was the headwear of the majority of Russian troops throughout the war. However France, as she had done with her other allies, agreed to supply Russia with helmets.

The Russian Adrian Model 1915-16 Helmet

Save for the badge, this helmet was identical to the French Adrian Model 1915. It was painted a chestnut brown to match the Russian uniforms. Different shades of brown will be encountered. The badge was made of pressed steel. It was 72mm high and 65mm wide and was painted in the same colour as the helmet. It was based on the cap badge worn previously by Russian officers and featured the double-headed eagle of Imperial Russia.

The Russians ordered some 2,048,000 helmets. It appears that France was able to supply only 340,000 by the end of 1916. The helmets were used initially on the Western Front, although they were also used in Macedonia and Salonica and on the Eastern Front.

Following representations made on behalf of the Russian military, on the 6th July 1916 a letter was written on behalf of the French government confirming that there was no objection to supplying the Russians with the necessary details to enable them to manufacture the helmets for themselves. Thereafter, the helmets may have been manufactured in Russia although details of the quantities produced, if any, are unknown.

Variations of the standard helmet have been observed. These include blue-painted helmets with the Russian badge in blue, blue-painted helmets with the Russian badge in brown, and brown-painted helmets with the Russian badge in blue. It has been said that the blue painted helmets were the first to be supplied before brown painted helmets became available. This may be correct, but it has not been possible to find definite confirmation. It is possible that the blue-painted helmets with a brown badge were a field replacement of a damaged brown helmet with a French blue helmet with the French badge being substituted for the original Russian badge. It does seem unlikely that a helmet would have left the manufacturer with a badge of a different colour to the helmet, although it is possible.

Given the number of Russian Adrian Model 1915-16 helmets produced, it might be expected that the helmets would be relatively common today, but that is not the case. One of the reasons is that following the abdication of the Tsar and the subsequent Treaty of Brest-Litovsk on the 15th December 1917 ending the war between Russia and the Central Powers, Russia was involved in a number of conflicts. These were both internal and external. The Russian Adrian Model 1915-16 Imperial helmets were used, usually with the Imperial badge removed. Other countries, apart from Russia, made use of the Russian helmets. Again, this usually involved the removal of the original badge.

The Russian "Sohlberg" Model 1917 Helmet

In 1916 the Russians ordered a number of these helmets from Finland, then part of the Russian Empire. They were to be produced by the Sohlberg factory at Helsinki. In addition, helmets were manufactured in Russia. The Tsar had intended this helmet to be the new Russian helmet.

It has not been possible to examine a Russian Sohlberg helmet in original state. The only helmet examined appears to be one captured from the Russians by the Austro-Hungarians and converted by the removal and replacement of the liner and chinstrap and chinstrap attachments. However, it is possible to comment that the shell of that helmet measures 294mm from front to rear and 201mm from side to side. The rivets used to attach the metal strips securing the liner headband have been removed and the holes filled with metal plugs. However, the holes were only approximately 36mm apart and were therefore closer together than those on the Finnish helmets.

The circular ventilation hole at the apex of the helmet was 12mm in diameter. The ventilation hole cover was 46mm in diameter and was of a different shape to that found on the Finnish helmets. The cover was not star-shaped, since the "points" of the star were rounded, making the fitting circular in shape. There were larger semi-circular cut-outs in the sides of the dome to allow for ventilation. The ventilation hole cover was fixed to the helmet by three dome-headed split pins rather than rivets as found on the Finnish helmets.

Further details of the Sohlberg helmet may be found in the sections on Finland and Austro-Hungary.

Russian Adrian Model 1915-16 helmet. (Collection of Ellwood von Seibold)

Russian Adrian Model 1915-16 helmet. (Collection of Marcus Cotton)

Russian Adrian Model 1915-16 helmet badges.

Interiors of Russian Adrian Model 1915-16 helmets. Left: (Collection of Brana Radovic). Right: (Collection of Polish Army Museum)

12079MWP

Top: Russian Adrian Model 1915-16 helmet with blue-painted finish. Badge absent. (Collection of David Acon). Bottom: Blue-painted Russian Adrian Model 1915-16 helmet badge. (Collection of Brana Radovic)

Russian infantryman wearing a Russian Adrian Model 1915-16 helmet

Russian troops wearing Russian Adrian Model 1915-16 helmets. Champagne, France. Image supplied by Nik Cornish at Stavka Military Image Research

■ Top: Russian prisoners wearing steel helmets drawing a bread ration. These prisoners were captured (1,300) during an attack by General von Mackensen's Army near Focasani, 6th August 1917. IWM Q60341

■ Bottom: Russians advancing to the fighting line. Each man has one of the new helmets. IWM Q32401

Russians troops halted on the roadside: note each man has a French pattern steel helmet, Balkan Front, September 1916. IWM Q32402

SERBIA

CHAPTER SEVENTEEN

SERBIA

Well before 1914 Serbia was an independent Slav state. Other Slavic people living within the Austro-Hungarian Empire were anxious to link up with her. Russia saw herself not only as a supporter of Serbia, but also as the champion of all Slavs living within the Austro-Hungarian Empire. Slav unification could either take place within, and as a part of the Austro-Hungarian Empire, or by the expansion of Serbia at the expense of the Empire.

In 1908 Austria annexed the former Turkish province of Bosnia-Herzegovina. Serbia was a land-locked country with an ambition to acquire an outlet on to the Adriatic. The annexation of Bosnia-Herzegovina not only frustrated that ambition but also provided Austria with a suitable base from which to launch an attack on Serbia. Tension developed between Serbia and Austro-Hungary on the one hand, and between Austro-Hungary and Russia on the other.

Germany was closely tied to and supportive of Austro-Hungary, her neighbour to the south. At the same time Germany resented the alliance between Russia and France, her neighbours to the east and west respectively. The Anglo-Russian agreement in 1907 gave rise to further anxiety in Germany, causing a fear of encirclement.

Further fuel on the fire that was to become the Great War was provided by the Serbian victory against Turkey in the First Balkan War in 1912. German aspirations were best served by maintaining the status quo and stability in Austro-Hungary. The position of supremacy of Austro-Hungary in the Balkans was threatened. The threat to Austro-Hungary, coupled with the enhancement of Russian status as supporter of the Slavs and Serbia, increased animosity within Germany towards Russia and Serbia. The position that developed was one of the needs and aspirations of the German people against those of the Slavs, including Serbia.

Following the Balkan War, Austrian pressure was exerted to persuade Turkey to agree to the creation of an independent Albania. At the same time Greece, under German influence, annexed Thrace from Turkey. The combined effect was on the one hand to cut Serbia off from the Adriatic and, on the other hand, to deny her access to the Aegean.

In the Second Balkan War Serbia occupied Albania, thus obtaining an outlet to the Adriatic. Following an ultimatum from Austria she withdrew, but a serious atmosphere of instability remained. Russia continued to offer support to Slavs, both within the Austro-Hungarian Empire and within Serbia. Austro-Hungary regarded Serbia as being responsible for fermenting antipathy towards the Austro-Hungarian Empire.

On the 28th June 1914, the Archduke Franz Ferdinand, heir to the Austro-Hungarian Empire, and his wife were assassinated at Sarajevo by a Bosnian Serb. Within the Austro-Hungarian Empire there was uncertainty as to the appropriate response. Retribution against Serbia had to be weighed in the balance with the fear that such retribution might bring Russia to Serbia's assistance. German support for Austro-Hungary, coupled with a belief that Russia was unprepared for war, served to tilt the balance.

Terms of an ultimatum to be delivered to Serbia, involving fifteen demands, were finalised on the 19th July 1914. Despite the fact that it was clear that Serbia would reject the terms of the ultimatum, on the 23rd July 1914 it was delivered. A response was required within forty-eight hours. On the 25th July 1914 Serbia mobilised her army. She also delivered what was regarded as a conciliatory response to the ultimatum. Despite pressure against war, on the 28th July 1914 Austro-Hungary declared war on Serbia.

Strenuous efforts to avert an escalation of the conflict followed but, on the 30th July 1914, Russia mobilised. Confident of German support, and faced with the Russian mobilisation, on the 31st July 1914 Austro-Hungary also mobilised.

Germany was hopeful of a quick victory against ill-prepared Russia. However, a war on two fronts was not desirable. Germany was anxious to obtain the benefits of a successful war in the east with Russia whilst, at the same time, fearful of a successful attack in the west by France in support of Russia. Thus, before declaring war, Germany sought an assurance from France that she would remain neutral. That was rejected and France mobilised.

The answer to this dilemma was an immediate attack on France to swiftly overcome her, leaving the way clear to start the necessarily longer campaign to overcome Russia. In 1905 the Chief of the German General Staff, Alfred von Schlieffen, had finalised the detail for the attack on France. The Schlieffen Plan, as it was known, involved a sweep by German forces through neutral territory to enter France from the North and to capture Paris from the West, thus avoiding the French fortifications facing the Franco-German border. Despite continuing efforts to avert war, on the 1st August 1914 Germany declared war on Russia and on the 3rd August 1914 against France. On the 4th August 1914 Britain declared war against Germany.

On the 12th August 1914, Austro-Hungary invaded Serbia. Initially, the invasion met with success. However, fierce resistance and a counter-attack by the Serbian Army lead to the Battle of Jadar. The Austro-Hungarians were pushed back. Austro-Hungary persisted and on the 1st December 1914 took Belgrade. Although the Serbs managed to re-occupy Belgrade, they were driven out again on the 9th October 1915 during the Austro-German offensive. Two days later the Bulgarians joined in the attack against Serbia.

Although the Allies sought to assist Serbia, her forces were driven back and were obliged to retreat into Albania. Serbian soldiers and civilians alike were pushed back. They retreated across the mountainous, harsh terrain suffering heavy casualties, both in battle and from disease and deprivation, as they went.

Following the tragic retreat of the Serbian Army at the end of 1915, some 260,000 survivors were transported to Corfu by Italian, French and British ships. The British and French forces that had endeavoured to support Serbia were obliged to retreat into Greece. The troops of the Central Powers occupying Serbia, mostly Bulgarians, entrenched along the former Serbian border. For the time being, Serbia was left to her fate.

Almost immediately, on the 8th January 1916, Austro-Hungary attacked Montenegro, a country that was Serbia's neighbour and ally. Montenegro was forced to surrender after only nine days, although some Montenegrin troops escaped and later joined the Serbian forces.

The Serbian army was re-equipped by the French and British. During April 1916, 125,000 Serbian soldiers left Corfu for the Salonica Front where they reinforced the British and French forces. The Serbs took part in the offensive launched on the 12th September 1916. The Bulgarians, who had been reinforced by German troops, were pushed back and on the 19th November 1916 the Serbian town of Monastir was re-taken. It is interesting to note that Monastir was only taken from the Turks by Serbia in the First Balkan War in 1912.

Serbian troops continued to fight with the Allies on the Salonica Front. A new offensive was launched on the 8th May 1917 and gains were made, only to be lost. It proved impossible to breach the Bulgarian defences.

In his speech to the United States Congress on the 8th January 1918, President Wilson included the principle that the peoples of the Austro-Hungarian Empire should be given the opportunity of autonomous development. Austro-Hungary would vacate Serbia. Serbia would be restored and given access to the sea. A general association of nations would be formed, he proposed, to guarantee the political independence and territorial integrity of great and small nations alike.

A further offensive on the Salonica Front was launched on the 14th September 1918. It was more successful and, despite heavy resistance, the Allies advanced. What had been the Salonica Front was pushed forwards into Serbia. A Yugoslav Division had been formed. This represented a determination by the Slovenes, Croats, Serbs, Bosnians, Montenegrins and Macedonians to unite, at least territorially, once the Austro-Hungarian forces had been driven out. Following ferocious fighting the determined resistance by the Bulgarians weakened and, with mutiny in two regiments, it became clear they many would fight no more. Although stubborn resistance by some troops continued to the last, on the 20th September 1918 the Bulgarian army was ordered to retreat.

Although Austro-Hungarian troops continued to resist as Serbian, French and British forces pushed forwards the resistance was no longer as effective. In an effort to preserve the Austro-Hungarian Empire, the Emperor Charles offered federal freedom to all six principal nationalities of the Empire, including the Serbs. The offer was not accepted, President Wilson insisting that autonomy within the Empire was inadequate. On the 25th October 1918, Count Michael Karolyi set up the Hungarian National Council in Budapest. This was the prelude to the separation of Austria and Hungary and the complete collapse of the Hapsburg Empire. On the 31st October 1918, Serbian troops reached the heights above Belgrade and were able to watch the Austro-Hungarian forces retreating back across the Danube.

As mentioned, the British and French re-equipped the Serbian army, their helmets being supplied by the French.

The Serbian Adrian Model 1915-16 Helmet

The helmet, with the exception of the badge, was identical to the French Adrian Model 1915 helmet. Please refer to the section under France for more detail. The vast majority of the Serbian helmets were painted light brown to match the khaki uniforms supplied by the British. A few helmets, used by officers who retained their grey uniforms, were painted artillery grey or grey-green.

The badge used on the Serbian Adrian Model 1915-16 helmets was that of the Karageorgevic Monarchy. The badge consisted of a double-headed eagle with outstretched wings with the royal crown above. On the eagle's breast was a shield featuring a cross and four letter "Cs" two facing right and two facing left. The badge measured 73mm high and 54mm wide.

Although a larger number of helmets were ordered, France supplied only approximately 123,000. The helmets were first used in combat in September 1916 against the Bulgarians.

It is worth mentioning that the Serbian army, and subsequently that of Yugoslavia following the creation of that state in 1929, continued to use the Adrian helmets up to the Second World War when, in 1941, the German Army overran Yugoslavia. On the creation of Yugoslavia a new badge was adopted. The new badge was the same as the original Serbian badge save that the shield bore the emblems of the Serbs, Croats, Slovenes and Bosnians.

It is possible that the Bulgarians made use of captured Serbian helmets during the Great War. Certainly, Serbian helmets converted by the Bulgarians exist. It is not known for certain whether these helmets were captured and used during the Great War or whether they were captured in the Second World War. For further details, please see under Bulgaria.

Serbian Adrian Model 1915-16 helmet. (Collection of Marcus Cotton)

Serbian Adrian Model 1915-16 helmet. Note the battle damage caused by a bullet. (Collection of Brana Radovic)

Serbian Adrian Model 1915-16 helmet. Note the replacement chinstrap. It is possible that this was an officer's helmet. The chinstrap appears to be made from leather taken from another piece of equipment. (Collection of Ellwood von Seibold)

Serbian Adrian Model 1915-16 helmet. The badge is absent but it is obvious from the shape left under the badge that this is a Serbian helmet. It is to be noted that manufacturers' procedures varied. Some Adrian helmets were painted with the badge in place whilst others were painted before the badge was fitted. Opposite: the interior of the above helmet. Note that the leather of the liner has shrunk, possibly due to exposure to gas. (Collection of Michael Haselgrove)

Firing the trench mortars. A Frenchman is instructing the Serbians. The shell weighs about 100lbs. Note the Serbian Adrian Model 1915-16 helmet worn by the soldier on the right. IWM Q32146

■ Top: Battle of Kamachalan-Florina. Serbian soldiers resting behind the line after many days fighting. Balkan Front, September, 1916. Note that they wear the Serbian Adrian Model 1915-16 helmet. IWM Q32392

■ Bottom: Battle of Kamachalan-Florina. Serbian troops outside their shelters in a front line trench, Balkan Front, September 1916. Note the Serbian Adrian Model 1915-16 helmet. IWM Q32327

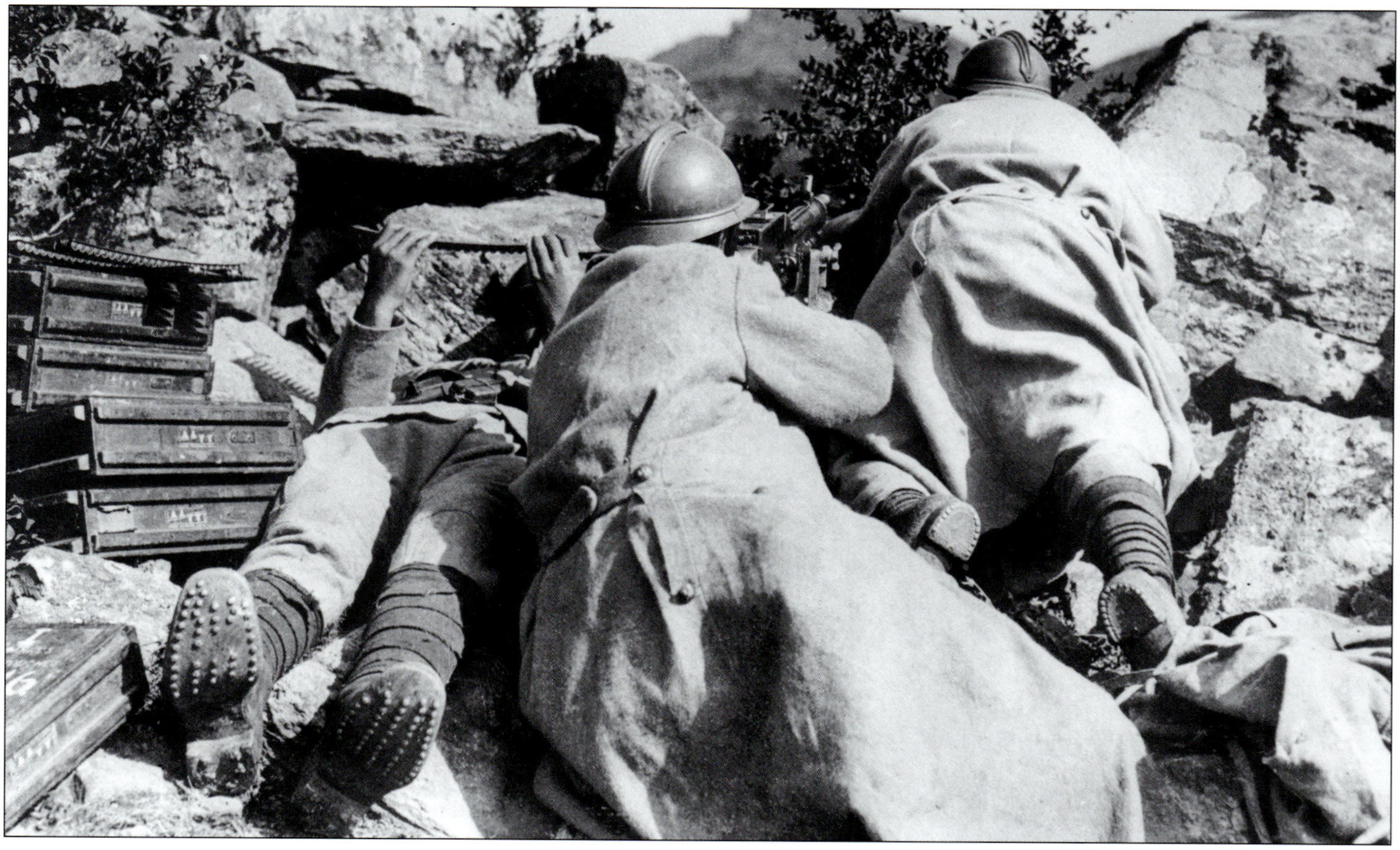

Top: Serbian officers examining their new trench mortars. IWM Q32143

Bottom: Mountain fighting. Machine gun behind rocks, 600 yards from the Bulgars. Thought to be Battle of Doiran 1916. IWM Q32215

TURKEY

CHAPTER EIGHTEEN

TURKEY

In 1914 there was a crisis in the relations between Turkey and Greece, albeit somewhat obscured by events elsewhere. This arose in part as a legacy of the Balkan Wars of 1912 and 1913 when the Balkan Allies, including Greece, had sided against Turkey. A race for naval supremacy had developed, Turkey ordering two battleships from Britain and Greece ordering two cruisers from the United States of America.

The coming of the Great War prevented the outbreak of war between Turkey and Greece, but it was no surprise that the sympathies and interests of the two countries were seen to be on opposite sides. Turkey had been for some time, if not an actual member of the Triple alliance with Germany, Austro-Hungary and Italy, then a nominal member. Turkey was, if anything, a closer ally to Germany than Italy, having developed an increasingly close relationship with Germany. Germany sought to extend her sphere of influence, supporting Turkey financially and with military expertise and equipment. It was obvious once the war started that Turkish neutrality would not continue for long.

At the beginning of August 1914, the British informed the Turkish Government that they were taking over the Turkish battleships ordered and under construction in Britain. Later that month Germany sold the German warships Goeben and Breslau to the Turkish Government. The Allies sought to keep Turkey out of the War by offering to uphold Turkish independence and integrity against any enemy if she would remain neutral. However, German influence over Turkey was too great and, despite diplomatic activity, on the 2nd November 1914 she entered the war on the side of the Central Powers.

In the years before the Great War the Turkish Army had been the subject of reform and modernisation. Brightly coloured uniforms had given way to khaki. Germany had supplied much modern military equipment, including small arms and artillery. The Balkan Wars, and religious differences existing within the Turkish Empire, hampered full implementation of reforms. However, there had been considerable progress with modernisation by 1914. The belief held by the Allies that Turkish resistance in the Dardanelles would be overcome easily was quickly proved to be misplaced.

The Turkish Model 1918 helmet

The Turks had experimented with protective helmets before August 1914. The Balkan Wars had shown the desirability of such a helmet. However, the experiments were on a modest scale and when Turkey entered the Great War the headwear worn by the majority of her troops was the Kabala. The Kabala was a hat formed by winding a strip of cloth around a semi-rigid frame. Save against the weather it provided little protection.

Thus, for a considerable time Turkish troops did not have an effective helmet and indeed most troops did not receive a steel helmet before the end of the war. Even when helmets became available and were supplied to Turkey by Germany, the number issued appears to have been modest. The reasons for this were firstly the inertia of the government and secondly that Turkish troops were operating in zones where the climate was hot, making the wearing of a steel helmet impractical.

Additionally, some attribute the delay to the difficulty in designing a helmet that was suitable and in compliance with religious requirements. The helmet designs adopted by other nations were unsuitable, it is said, in that they all had a brim or visor preventing religious prostration.

It was only later in the war that Germany agreed to supply Turkey with appropriate helmets. There were three types of helmet provided, all derived from, or based upon, the standard German helmets. It appears appropriate that all three versions of Turkish helmet be designated the Turkish Model 1918 Helmet. Manufacture of all three versions of the Turkish Model 1918 Helmet was very limited. Indeed, in respect of one of the three types only 5,400 examples are thought to have been produced.

The first type was a cut-down version of the standard German Model 1916 or Model 1917 helmet. The conversion, believed to have been carried out locally, involved the removal by cutting of the visor and the lower brim. The converted helmets had an approximately level base. The lower edge was unrolled and unfinished in any other way. The example examined was marked ET66. The helmet base measured 287mm from front to rear and 224mm from side to side. The measurement from the bottom of the chinstrap attachment rivet head on the outside of the helmet to the base of the helmet was 7mm on each side. That measurement on a standard size 66 German helmet was 22mm.

The Germans supplied the second type of helmet in Spring 1918, manufacturing them without a brim. The front of the helmet was simply a continuation of the rear, without the familiar step on each side or the projecting brim of the standard German helmet. These Turkish helmets had the entire lower edge folded inwards and were manufactured for Turkey specifically and were not a conversion of existing standard German helmets. Now often referred to as the full-visor model, the intention was to issue them to elite infantry units. The example of this helmet examined was again marked ET66 and measured 290mm from front to rear and 232mm from side to side.

The third model, sometimes referred to as the visorless model, was without the front brim and visor and had a very different profile to the previous helmets. With this helmet, the front curves across the wearer's forehead, leaving the forehead partially exposed. Again, the lower edge of the helmet was folded inwards to produce a smooth finish. It is probable that this pattern of helmet never reached Turkey. Photographic evidence suggests that the helmets were issued to Freikorps troops in Germany following the end of the Great War. The helmet examined was size 66 and measured 282mm from front to rear and 237mm from side to side. At the front centre the distance between the bottom of the liner band and the base of the helmet was 19mm.

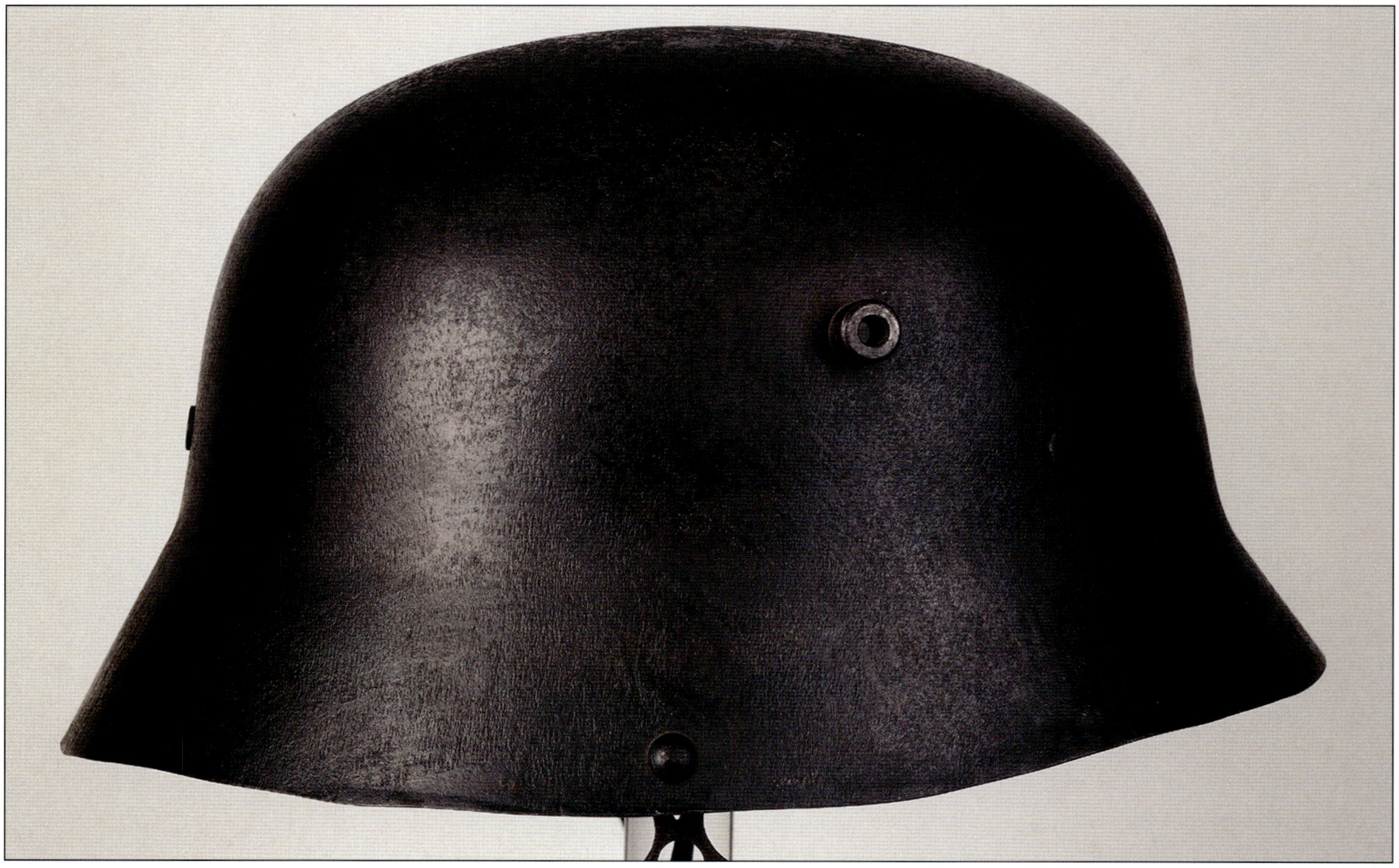

Turkish Model 1918 Helmet (cut-down version). (Collection of Brana Radovic)

Liner of helmet shown on previous page. Note German Model 1916 liner. (Collection of Brana Radovic)

Turkish Model 1918 Helmet. (this page and over) Note camouflage paint finish, the German Model 1917 liner and battle damage. (Collection of the Imperial War Museum)

Turkish Model 1918 Helmet (full visor version). Note German Model 1918 liner and that the chinstrap is not the standard Model 1918 type. (Collection of Marcus Cotton)

Turkish Model 1918 Helmet (visor-less version). (Collection of Brana Radovic)

Top and bottom: This officer and the soldiers below all wear the German-supplied Turkish Model 1918 helmet. (Bottom photograph: IWM Q80044)

UNITED STATES

CHAPTER NINETEEN

UNITED STATES

Prior to April 1917, the United States of America remained neutral. That is not to say she did not play a part. She provided a huge amount of material support, particularly to the British. This ranged from small arms ammunition to submarines.

An early source of friction between the United States and Germany was the commencement, announced on the 4th February 1915, of submarine warfare against all shipping bringing supplies to the Allies. This included neutral shipping and, on the 28th March 1915, the first American citizen to die in the Great War was lost when the German submarine U-28 sank the Falaba. The sinking of the passenger liner Lusitania on the 7th May 1915 by the U-20 with the loss of 128 American lives shocked America. However, it did not bring her into the war.

What can be said is that the sinking of the Lusitania and other shipping, coupled with the execution by the Germans of a British nurse, Edith Cavell, on the 12th October 1915 did turn American public opinion against Germany. However, in May 1916 Germany agreed to cease unrestricted submarine warfare, although the right was reserved to stop and search shipping and to sink any ship carrying war material. Despite this, the loss of shipping to U-boats continued, albeit on a reduced scale.

President Woodrow Wilson was re-elected on the 7th November 1916. He continued to seek peace, writing to the countries involved in the war seeking a resolution of the conflict. It seemed possible to some that his initiative might be successful but, in reality, the war had developed a momentum of its own. The armies of the combatant nations were expanding and, in that context, any proposals to end the war put forwards by one side were unsatisfactory to the other. The prospect of defeat was required before there could be any real hope of peace.

The entry of Romania into the war on the side of the Allies in August 1916, coupled with the pressure caused by the Russian offensive in the east and the Franco-British offensive on the Somme, caused the Kaiser to accept advice to resume unrestricted submarine warfare in early 1917. This included the sinking of American shipping, meaning that war was virtually thrust upon the American nation. Germany was prepared to accept the risk, having concluded that unrestricted submarine warfare would force Britain out of the war within six months.

On the 23rd January 1917 President Wilson announced to Congress that he was breaking off diplomatic relations with Germany. This was poor news for Germany. However, it was to some extend off-set by the hope of British defeat and the worsening military and political position in Russia.

Another factor serving to bring America into the war was a scheme by Germany to enter into a pact with Mexico. That country was to be encouraged to think in terms of recovery from America of territories lost many years previously, namely New Mexico, Arizona and Texas. Developed by Dr. Alfred von Zimmermann, a German Foreign Minister, the scheme was a closely guarded secret. Unfortunately a telegram sent by Zimmermann was deciphered and made public in March 1917. Although the scheme may today appear fantastic, at the time it was taken seriously and, coupled with the resumption of unrestricted submarine warfare, served to sway public opinion in favour of entry into the war.

Further provocation of America by Germany came with the sinking of more American merchant ships with loss of American lives. On the 6th April 1917 America declared war on Germany. At that time, the American Army was relatively small. An army of at least one million, and possibly as many as three million men, would be required. The huge task of recruiting, equipping, training and transporting such an army would take a minimum of a year and, in reality, rather longer and it was clear that American troops would not have significant influence on the battlefield during that time.

On the 5th June 1917 the Selective Draft Act came into force. All males aged between twenty-one and thirty were required to register as a first step leading to military service. Whilst that process got under way, a huge programme was initiated to build the large number of merchant ships, required to transport and supply the American Expeditionary Force (AEF as it came to be known), and warships.

General John J. Pershing was offered and accepted the command of the AEF. He and his staff left New York for England on the 28th May 1917. The first large contingent of American troops arrived in France on the 27th June 1917. The troops were of the 1st Division AEF and they took considerable pride in the fact that they were first in a number of things, including first in sector on the battlefield and first to engage the enemy.

However, although the presence of American troops in France greatly boosted the morale of the Allies, some time would elapse before their presence was felt on the battlefield. Further training and reinforcement was required and artillery had to be provided. It was usual for troops inexperienced in trench warfare to enter the line in a quiet sector alongside experienced troops. In the case of the 1st Division, they first entered the line in the Sommerville (Lorraine) Sector on the 21st October 1917. They suffered their first casualties, the first Americans to die in combat in the Great War, on the 2nd November 1917. Later, the 1st Division was the first full American division to enter the front line, doing so in the Ansauville sector in the St. Mihiel Salient on the 15th January 1918.

Over a period of time, considerable pressure was exerted upon General Pershing by Britain and France to send troops immediately to serve with the British and French Armies. With the ending of the war between the Central Powers and Russia, releasing in particular large numbers of German troops to serve on the Western Front, there was great concern that the German offensive of early 1918 would overwhelm the British and French Armies and that the arrival of the AEF in force would be too late. Equally, Germany was anxious to launch the offensive before the build-up of American troops on the Western Front could prove decisive.

With the exception of an earlier agreement to release four regiments of black troops to serve with the French, General Pershing resisted the pressure. The AEF would serve as an independent army and not as part of the French and British Armies. As a result, there was considerable tension between, on the one hand British and French High Command and, on the other, the Americans. Eventually, on the 2nd May 1918, an agreement was reached that 130,000 Americans being transported that May and another 150,000 in June would join the British and French lines.

At the end of May 1918, at Cantigny on the Somme, the AEF launched their first offensive. Although the Americans held Cantigny, elsewhere the Germans were successful. The French and British were under great pressure and General Pershing agreed to vary the agreement. The new agreement provided for an increase in the number of Americans joining the British and French in June to 170,000, and to a further 140,000 in July. With those exceptions, troops transported to Europe joined the AEF.

Some of those troops released by General Pershing were sent to the Marne where they halted the German advance at Chateau-Thierry and Belleau Wood. Despite this success, General Pershing was under no illusion as to the pressure the French and British were under after nearly four years of war. The main thing in his opinion was to hold the line until the AEF was sufficiently strong to ensure victory. He requested the War Department in Washington to vastly increase the AEF, increasing it from 800,000 men in June 1918 to 2.5 million men by May 1919.

The tide was beginning to turn in favour of the Allies. The successes at Chateau Thierry and Belleau Wood were followed by further successes, in particular in the offensives at St. Mihiel and the Meuse-Argonne in September 1918. The beginning of October 1918 saw the Allies advance on all sectors of the Western Front. German resistance was particularly stubborn in some sectors, and few will not be familiar with the story of the Lost Battalion. The Germans were finally driven from the Argonne Forest on the 10th October 1918, having been in occupation since 1914.

The American Offensive on the Meuse, starting on the 1st November 1918, was a great success reaching the river opposite Sedan on the 6th November 1918. There were rumours of an armistice but the fighting continued with stubborn German resistance to American advances across the Meuse River. Despite the fact that some commanders knew several hours in advance that the Armistice was to come into effect at 11am on the 11th November 1918, orders were issued that planned attacks were to take place. This was in spite of the fact that there could be no possible advantage gained and that unnecessary casualties would be sustained.

Following the armistice, a force of 16,000 American soldiers was sent to Germany as part of the occupying forces. Other troops were sent home, the last division leaving France on the 1st September 1919.

It must be remembered that American troops were engaged elsewhere than on the Western Front. Some 135 Americans served in the American Red Cross in Italy. America had declared war on Austria on the 7th December 1917 and, although the need for troops on the Western Front was most pressing, an American regiment was sent to Italy. It arrived in Milan on the 28th July 1918, the 2nd Battalion entering the line in a quiet sector by the end of August 1918. The 1st and 3rd Battalions joined the 2nd Battalion in the lines near Treviso in October 1918.

Those troops of the 332nd Infantry regiment, together with several supply and medical units, were the only American troops sent to Italy. They were sent to the front on the Piave River with the intention that their presence would boost Italian, and lower Austrian, morale. They were ordered to give the impression, by moving around a lot and spreading themselves thin on the ground, that their force was much larger than was the case. By this stage the Austrians were largely a spent force and were retreating. The 332nd Regiment did engage the Austrians on the Tagliamento River, suffering one killed and seven wounded. The fighting in Italy ended on the 4th November 1918.

Other American troops were sent to Russia. They, and the troops of other nations involved, were intended to start a new front to divert German resources on the Western Front and to protect Allied stores at Murmansk and Archangel against the Germans.

The Allies had seized both ports and, at the beginning of August 1918, had driven the Bolsheviks out. Approximately 4,500 American troops arrived on the 11th September 1918 to support the Allied force and an outpost front of some 450 miles was developed. Approximately 174 Americans lost their lives before the American force was withdrawn in late 1919.

American Helmets during the Great War

Prior to her entry into the war, America had given consideration to the adoption of a suitable protective helmet for her armed forces. However, no such helmet had been adopted and, in June 1917, a committee was formed to consider the matter. Various designs, including the standard British and French helmets, were considered.

Time passed and the matter became urgent. It was felt there was insufficient time to design, test and manufacture a new model of helmet. In those circumstances the decision was taken to adopt the British helmet, despite its perceived shortcomings.

The decision to adopt the British helmet was influenced by the fact that the steel from which the British helmet was manufactured offered better protection than the French helmet. Additionally, America had earlier in the war supplied large quantities of war material and expertise to Great Britain and her allies. By 1917 Great Britain was largely self-sufficient as far as the manufacture of munitions and other war materiel were concerned. The British offered to supply some 400,000 helmets immediately to equip the American forces bound shortly for France. That offer was accepted. The helmets were supplied to American troops in France between July and November 1917, thus providing time for the Americans to manufacture and supply their own helmets to the British design. The American manufactured helmets were known as the Model 1917.

The adoption of the British helmet was intended as a temporary measure only. The intention was to adopt a helmet of superior design, thus giving the American troops a distinctive helmet. To this end, attempts were made not only to improve the design and protective qualities of the British helmet, but also to design and manufacture a completely new helmet.

The French Adrian Model 1915 Helmet

American personnel used this helmet to a limited extent during the Great War. Prior to April 1917, some American citizens were anxious to be involved in the conflict. Some volunteered to serve in the armies of other nations such as Canada and France. These soldiers were obviously equipped with the helmet of the nation or unit in which they were serving. Others volunteered to serve in a medical capacity in France or Italy. These personnel were supplied, if serving in a forward area, with Adrian helmets. The helmets were supplied without badges or the holes for fixing a badge. However, some of the helmets supplied to personnel of the U.S. Ambulance Corps were fitted with a distinctive badge consisting of a shield bearing the American flag surrounded by a laurel wreath (see page 667).

Shortly after April 1917, American personnel arrived in France. American helmets were not available and French Adrian Model 1915 helmets were worn. There is, for instance, a photograph of Major-General James G. Harbord (commander of the Marine Brigade at Chateau Thierry) wearing an Adrian Model 1915 helmet without holes for fixing a badge but with one rank star and a plaited two-piece chinstrap. This use very much mirrors the use by British officers of standard and privately purchased Adrian helmets earlier in the war before standard British helmets became available.

The American 93rd Division, which consisted of only four infantry regiments, namely the 369th to 372nd inclusive, also wore standard French Adrian Model 1915 helmets. These helmets, although supplied with the holes for fixing a badge, had no badge. This division never operated as a unit, the regiments being attached to the French army for the duration of the war. Although the supply of those helmets instead of the American Model 1917 to the 93rd Division (which consisted of black troops) may be seen as discriminatory, in fact the Adrian helmets were probably issued so that the American troops would wear the same helmet as the French troops they served with. In any event, the wearing of the French helmets became a source of pride. When divisional signs were adopted, the 93rd Division adopted a badge consisting of a disc with an Adrian helmet superimposed.

The American Model 1917 Helmet

Design, Introduction and Production

Save for a number of relatively minor differences, the helmet is the same as the standard British Mark I helmet. Reference should be made in the section under Great Britain for more detail.

The manufacturing detail of the American helmet does differ from its British counterpart in a number of ways. It is not proposed to give detail of the differences between the Model 1917 and the earlier British Helmets. These are fairly obvious. Having said that, care must be taken when considering any given helmet in that some helmets were equipped with a replacement liner and refurbished.

The basic differences between the post-1916 Mark I British helmet and the Model 1917 are as follows:

1. The chinstrap attachments on the British helmets are secured with a split pin whereas the American helmets have a dome-headed rivet. This is perhaps the most obvious difference. Having said that, at least one Model 1917 helmet has been observed where the rivet has a flat head on the interior rather than a dome or round-headed one. The important difference is that the American helmets have a rivet whilst the British helmets have a split pin.

2. The metal wire used to construct the rectangular chinstrap attachments and wire joints, joining the two sections of the

chinstrap assembly, is much thicker on the American helmets. On the American helmets the wire is approximately 2.8mm whereas, on the British helmets, it is approximately 2.3mm.

3. The metal rim fixed to the brim of the American helmets meets either flush at the ends or overlaps very little. On the British helmets there is normally a much larger overlap, often 25mm or more. This gives the British helmets a much less neat-looking finish. It also appears from the helmets observed that the metal rim is far better fitted on the American helmets. Often, on the British helmets, there are wrinkles in the metal rim caused by bending the metal to fit the brim.

4. Different stamping is applied to the inside brim of the respective helmets.

5. Different stamps are applied to the underside of the material forming the headband.

6. There are variations in the paint finish. Usually, the texture to the exterior of British helmets is achieved by the use of sand, although sawdust was sometimes used. On American helmets, sawdust was used. Additionally, generally speaking, the paint finish on American-made helmets is more evenly applied compared with British helmets. In particular, on British helmets it is often possible to see the marks left by a paintbrush particularly to the interior of the brim.

7. On British helmets made after early 1917, a rubber ring or doughnut was fitted to the apex of the helmet under the liner. This is not present on the Model 1917.

8. The label fixed under the washer at the interior apex of the helmet bowl, bearing the words *"Tighten cord and adjust net to fit the head"* is made of linen-backed paper on the Model 1917. On the British helmets ordinary paper, which was not as durable, was used.

Production of the Model 1917 began in 1917. The helmet shells and liners were manufactured separately. They were then brought together, at the factory provided by the Ford Manufacturing Company at Philadelphia, for painting, assembly, packing and shipping. Initially, helmets were packed and shipped overseas for issue to troops. By the summer of 1918, helmets were issued to the troops before departure abroad.

Helmet shells were manufactured by the following:

1. The American Can Company.
2. The Budd Manufacturing Company
3. The Crosby Company of Buffalo, New York
4. The Sparks Withington Company.
5. The Taylor-Wharton Company, Pittsburgh.
6. The Worcester Pressed Steel Company of Worcester, Massachusetts.
7. The Benjamin Electric Company of Des Plaines, Illinois.
8. The Bossett Corporation of Utica, New York.
9. The Columbian Enamelling and Stamping Company of Terre Haute, Indiana.

Liners were manufactured by the following:

1. The Leatherwear Company of America of New York
2. The Progressive Knitting Works of New York.
3. The Taylor Company of Buffalo, New York.
4. The Curtain Supply Company of Chicago.

The Manufacture of the American Model 1917 Helmet

Initially, manufacturers experimented with different steel alloys. These experiments were, to some extent, a repeat of those carried out by the British. The steel used, of necessity, had to be of adequate strength whilst retaining the quality of reasonable ease of manufacture. Speed and economy in the manufacturing process were of utmost importance. There was no point in using an alloy requiring many separate pressing operations, with many failures, to make the helmet shell. For example, the Crosby Company of Buffalo produced a helmet shell in chrome-nickel steel. This required five separate pressing operations together with annealing (heat treatment) to produce a helmet shell.

In the circumstances, it was decided to procure a supply of steel of the type used for British helmets. The British supplied some 4,000 sheets of manganese steel. Experience showed that this steel could be pressed in a single "draw" to form a perfect helmet shell with minimum wastage. This type of steel was used for all helmets thereafter, with the exception of 200,000 helmets produced by the Columbian Enamelling and Stamping Company of Terre Haute, Indiana.

The helmet shells were pressed from 50.8cm square sheets of steel. Once pressed, the excess metal was trimmed and the rim attached by welding. The chinstrap loops were attached by riveting. The helmet shells were stamped under the brim with a number. This was a reference to the temperature control number of the batch of steel. Thus, if there was a problem with one helmet in any given batch, the rest of the batch could be identified. The steel sheets used were tested by means of a punch strike in the corner. The steel passed if the punch produced a hole of the required depth without splitting the metal. A fixed percentage of finished helmet shells from each batch were subsequently subjected to a ballistic test.

The Liner and Chinstrap

The liner and chinstrap were identical to the British helmet save in respect of the differences already noted. Although there appears to have been no uniform method of marking the liners, ink stamps often appear on the underside of the leather-cloth liner material. Often a manufacturer's stamp is found giving the name of the manufacturer and year of manufacture. Additionally, an ordnance department inspection stamp may be found.

The chinstrap, which was made of cowhide, was felt to be less than satisfactory. Plans were made to replace it with an olive-drab woven strap with a quick release buckle. However, this modification was not introduced during the Great War, although a number of the experimental helmets had the woven chinstrap fitted.

Helmet Sizes

In common with the British helmet, all American Model 1917 helmet shells were manufactured in the same size. A small sample of helmet shells measured shows remarkable uniformity. They measured 312mm front to rear and 288mm side to side with little variation. Although an average of British helmets measured in the same way would probably provide very similar figures, individually the British helmets varied by up to 15mm each for the same measurements. The somewhat uneven fitting of the rims to British helmets to some extent may account for the variations.

Not all helmets examined were marked with a size. Those marked had the size stamped in the leather of the upper part of the chinstrap to one side of the securing rivet at the apex of the helmet. In some cases the stamp was hidden beneath the leather-cloth liner material. The size stamps seen were either 6 7/8 or 7 3/8, but no doubt other sizes exist.

In common with the British helmet, all sizes of liner had twelve pieces of rubber tubing inserted in alternate pockets in the liner headband. The difference in size was achieved by cutting the liner headband in such a way as to allow additional space under the chinstrap attachments where no rubber tubing was inserted. For the small sizes there was one empty loop beneath each chinstrap attachment. For the larger size(s) there were two empty loops under each chinstrap attachment. In this way it was possible to adjust the size of the liner headband, whilst still allowing space for twelve rubber inserts in the alternate pockets of the liner headband.

The Paint Finish and Assembly

The helmet shells were dripped in olive-drab paint. Thereafter, and before the paint was dry, sawdust was applied in a thin layer to the exterior of the shell using compressed air. The helmet shells were then placed in a drying room where they remained for one hour subjected to a temperature of 200 degrees Fahrenheit. They were then again dipped in paint and the drying process was repeated. Once this was done, the liner was fitted to the helmet shell and riveted in place, the rivet passing through the centre of the crown of the helmet shell.

American Model 1917 helmets do exist where a textured finish using sand instead of sawdust has been applied. The scale of manufacture and other details are presently not known.

The Wilmer Eye Shield

This mask was the suggestion of an American Oculist, Colonel W. Holland Wilmer. It was constructed of soft, pressed steel and had a rectangular slot on each side to allow for normal, forwards vision. In addition, there was a smaller, oval-shaped hole beneath each rectangular vision slot to allow for downwards vision. The mask had a foam rubber cushion around the entire inside edge to provide a comfortable fit against the wearer's face.

The mask was attached to the helmet by means of two springs. On each side of the mask the spring was attached to a metal loop. This loop was, in turn, attached to an anchor point formed from the metal of the mask. The other end of each spring was attached to the helmet brim. For this purpose three holes were drilled in line just inside the helmet brim and adjacent to, or just forwards of, the chinstrap attachments. A metal loop was inserted into one of the three holes on each side and the spring was attached by means of a spring-loaded clip. The metal loop could be moved forwards or backwards into one of the three holes to achieve the optimum tension on the spring.

When not in use the mask was worn on the brow of the helmet where it was held tightly in place by the tension of the two springs. The eye shield weighed only seven ounces and did not upset the balance of the helmet.

A significant number, about 30,000, of the eye shields were ordered but were later rejected on the grounds that they were not easily kept in position. Model 1917 helmets will be encountered from time to time where the holes have been drilled on each side of the helmet to enable the eye shield to be attached.

Experimental Helmets

As mentioned, the initial intention was to adopt the Model 1917 as a short-term measure. The helmet was regarded as unsatisfactory in a number of respects. In particular, it was thought to provide inadequate protection to the side and back of the head and no protection to the face and eyes. The high centre of gravity meant that the helmet tended to tilt forwards or backwards. It was a common sight to see troops running with one hand on the top of the helmet to keep it in place.

With the intention of improving or replacing the helmet, numerous experiments were conducted throughout the war. Tests were conducted on the French Polack and Dunand designs, amongst many others. Some of the helmets considered were field-tested in France. However, the Model 1917 was popular with the troops and there was resistance to change. The war ended before any new helmet could be introduced, putting an end for the time being to further experimentation. Indeed, it was not until 1936 that the modifications constituting the Model 1917A1 were ordered.

A system of numbering the experimental helmets developed. Some of the more important helmets considered are referred to below using that system of numbering and with photographs of the helmets appearing in the photographic section following.

Experimental Helmet Model No.2 "Deep Salade"

This helmet was designed in June 1917. The liner system was broadly modelled on the German system, with three liner pads arranged on a metal headband. The helmet shell was deep, with angles designed to deflect projectiles. Initial attempts to manufacture the helmet were unsuccessful. However, in late 1918 Messrs Ford & Company produced about 2,000 helmets.

The helmets passed ballistic tests and were found satisfactory. The Commander-in-Chief of American Forces commented favourably on the helmet but it was not accepted into general service.

Experimental Helmet No.5

This helmet had much in common with the above model. However, in the Model No.5 the lower edges did not extend so far downwards, easing production and still affording much greater protection than the Model 1917 helmet, particularly at the sides and rear. The same liner system as on the No.2 helmet was used.

There is photographic evidence showing this helmet in use in France. Further evidence that the helmet was actually put into use is provided by the existence of souvenir painted helmets. However, the helmet was rejected ultimately because it was too similar to the German helmet and because it offered insufficient advantage over the Model 1917 helmet. A model 5A was also produced and there were several other variations.

Experimental Helmet No.7

This was designed as a sentinel's helmet and to be worn by machine gunners and others in static, dangerous positions. No more than fifty helmets of this model were manufactured. The helmet was produced in three weights, namely 13, 15 and 18 pounds. The heaviest weight of this helmet was said to resist a direct hit from a service rifle bullet at 200 yards, and at shorter range if struck at an angle.

Again, the liner was in the German style with three liner pads fixed to a metal frame. Despite the weight it appears it was possible to wear this helmet for a considerable period.

The sides of the helmet were hinged to enable the helmet to be opened when putting it on or taking it off. A small catch at the bottom front (broken in the example pictured) kept the helmet firmly shut when in use. There were two small holes in the front of the helmet, on each side of the wearer's nose, to assist with ventilation. Despite this, it is likely that wearing this helmet in even warm weather would be unbearably hot. The helmet did not attract favourable comment from those testing it in France.

Experimental Helmet No.8

This helmet was very similar to the No.5 helmet but with the addition of a visor. The visor was of solid construction and was said to afford better protection than other models of visor, whilst allowing good vision. Again the liner system was based on the German style. The helmet had a woven chinstrap as used on the No.5 helmet and as adopted for the Model 1917A1 helmet many years later.

The Ford Motor Company manufactured approximately 1,300 of these helmets in November 1918. Bashford Dean commented that the helmet did not appear to have been adequately tested. In any event, little interest was shown in the helmet. This was possibly because the war ended, terminating further helmet development for the time being.

The Liberty Bell Helmet

Developed just before the end of the war, this helmet was provisionally accepted as the replacement for the Model 1917. One advantage of the helmet was said to be that it could be worn reversed. Quite why this should be an advantage is not clear. In reality, the helmet had very little to recommend it over the Model 1917. The helmet was not a popular design with the troops and, despite the fact that a maximum of about 2,000 of the helmets were produced in France, ultimately it was not adopted.

Some of the helmets produced had an additional plate riveted to the front. Different liner systems were fitted, including one in the German style.

Painted and Other Badges and Souvenir Helmets

The use of divisional signs on uniforms and other equipment dates from May 1918, and their use gradually spread. By the end of the war, all divisions had an individual sign. Whilst the wearing of divisional signs on uniforms was authorised, there appears to have been no such authorisation for their use on helmets. Contemporary photographs, coupled with the recollection of veterans, indicates that all or virtually all marking of helmets by way of the fixing or painting of badges, or other painting such as camouflage finishes, dates from after the end of the war.

At the end of the war Congress agreed that each member of the armed forces might keep as a souvenir one complete uniform, including a helmet. The practice of marking the helmet with a divisional sign or otherwise painting the helmet as a souvenir appears to have been widespread. Some individuals painted their own helmet. However, in other cases, an individual or individuals were detailed to paint all the helmets of a unit in a uniform way. There is record of enterprising individuals setting up small "cottage" industries devoted to the painting of helmets in return for payment. In some cases, the original finish was rubbed down and the helmet repainted to provide a smoother surface for painting. Some helmets were painted with bright "camouflage" paint after the German style. Other helmets were painted with details of the owner's service, unit etc. Battle damage could even be inflicted on the helmet if the owner desired.

Some of the helmets were painted crudely. Others show great care and attention to detail and are of considerable artistic merit.

■ American Model 1917 Helmet. (Collection of Michael Haselgrove)

TIGHTEN COID
AND
ADJUST
NET
TO
IT THE HEAD

American Model 1917 Helmet with Wilmer Eye Shield. (Collection of Marcus Cotton)

American Model 1917 Helmet decorated as a souvenir, presumably post-war, with the initials AEF. (Collection of Ellwood von Seibold)

American Model 1917 Helmet decorated as a souvenir with the insignia of the 2nd Division AEF or "The Indian Head Division". The 2nd Division was organised partly in France and partly in America with units from America joining from time to time. Activities in France began on 15th March 1918 in the Verdun and Toul-Troyon sectors and ended in the Meuse-Argonne sector in November 1918. The red castle in the background indicates engineers. (Collection of Brian Maederer, Military Collectibles)

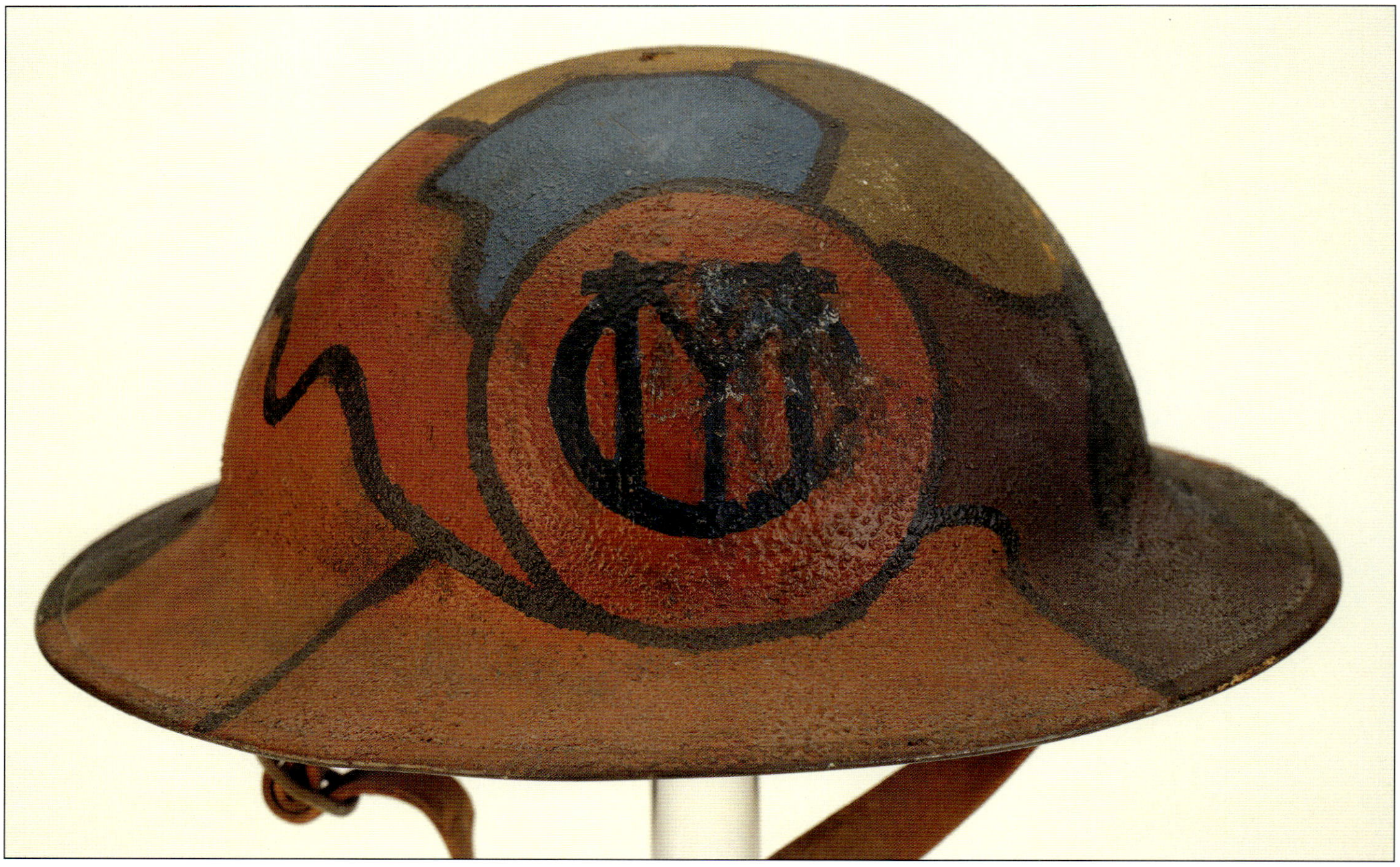

American Model 1917 Helmet. Superbly painted as a souvenir with a camouflage finish in the German style. The badge consists of initials, which appear to be "M Y", contained within a horseshoe. The unit is as yet unidentified. (Collection of Brian Maederer, Military Collectibles)

■ American Model 1917 Helmet. Camouflage painted souvenir helmet with the 90th Division insignia. Known as the "Texas and Oklahoma Division", it arrived in France on the 23rd June 1918. They saw action between August and November 1918 including the St. Mihiel Offensive between the 12th and 15th September 1918 and in the Meuse-Argonne between the 19th October and the 11th November 1918. (Collection of Brian Maederer, Military Collectibles)

American Model 1917 Helmet. Camouflage painted souvenir helmet with the badge of the 79th Division or "The Cross of Lorraine Division". The Division arrived in France on 15th July 1918. Their activities were in Sector 304 (between the Argonne Forest and the Meuse River) between the 16th and 30th September 1918 and the Troyon Sector between the 8th and 23rd October 1918 and the Grand Montaigne sectors (heights to the east of the Meuse) between the 29th October and the 11th November 1918. The struggle for Belleau Wood took place between the 6th and 25th June 1918 so the owner of this helmet was presumably in another unit at the time. (Collection of Brian Maederer, Military Collectibles)

BELLEAU-

American Model 1917 Helmet. Note sand finish and American cap badge attached by means of screw and nut passing through the helmet shell. (Collection of Marcus Cotton)

Various American Model 1917 helmets all decorated as souvenirs.

American Model 1917 Helmet fitted with a cloth drawstring cover that appears to have been made from an army-issue blanket.

Franco-American Experimental Helmet. Developed in France for the AEF, it carries a Dunand anti-splinter visor. On some helmets, where the visor is attached to the helmet shell on each side, a circular metal cockade with national colours is found. This example is without. Note the embossed U.S. national emblem to the front. On other examples the emblem was stencilled. The liner is leather and is of one piece construction with nine tongues with gold-embossed borders. At the end of each tongue is a hole, reinforced with a metal eyelet designed to take the drawstring used to adjust the liner. The leather liner is virtually identical to the liner fitted to the early Acier Trempe Adrian-style helmets. It may well have been made by the same companies, Siraudin & Co. and Franck & Co. (see section on France). The liner system is radical in that it is suspended from the helmet shell by four springs, spaced around the liner and presumably designed to absorb the shock of impact. Note that the chinstrap is friction adjustable, but has no buckle.

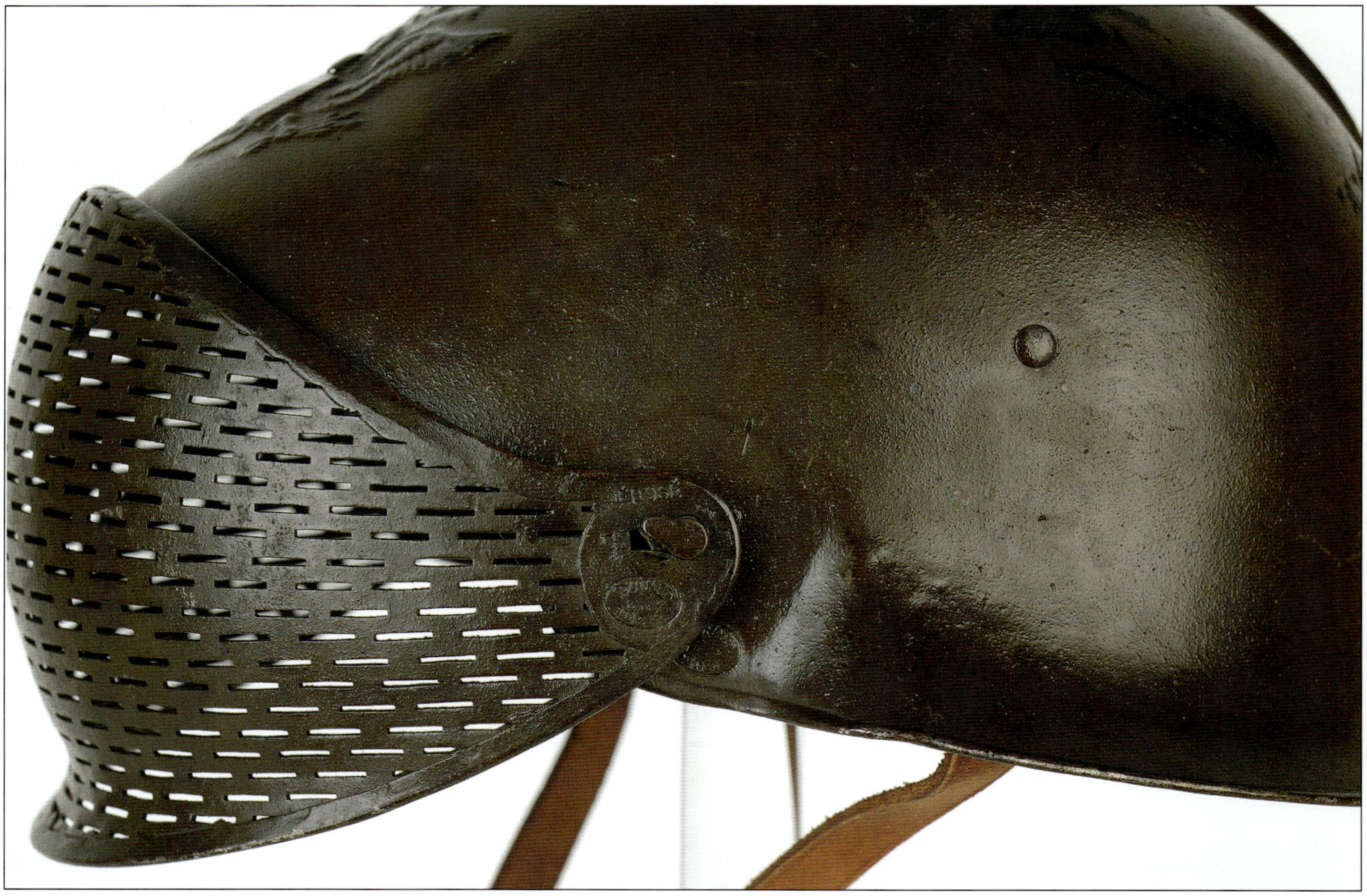

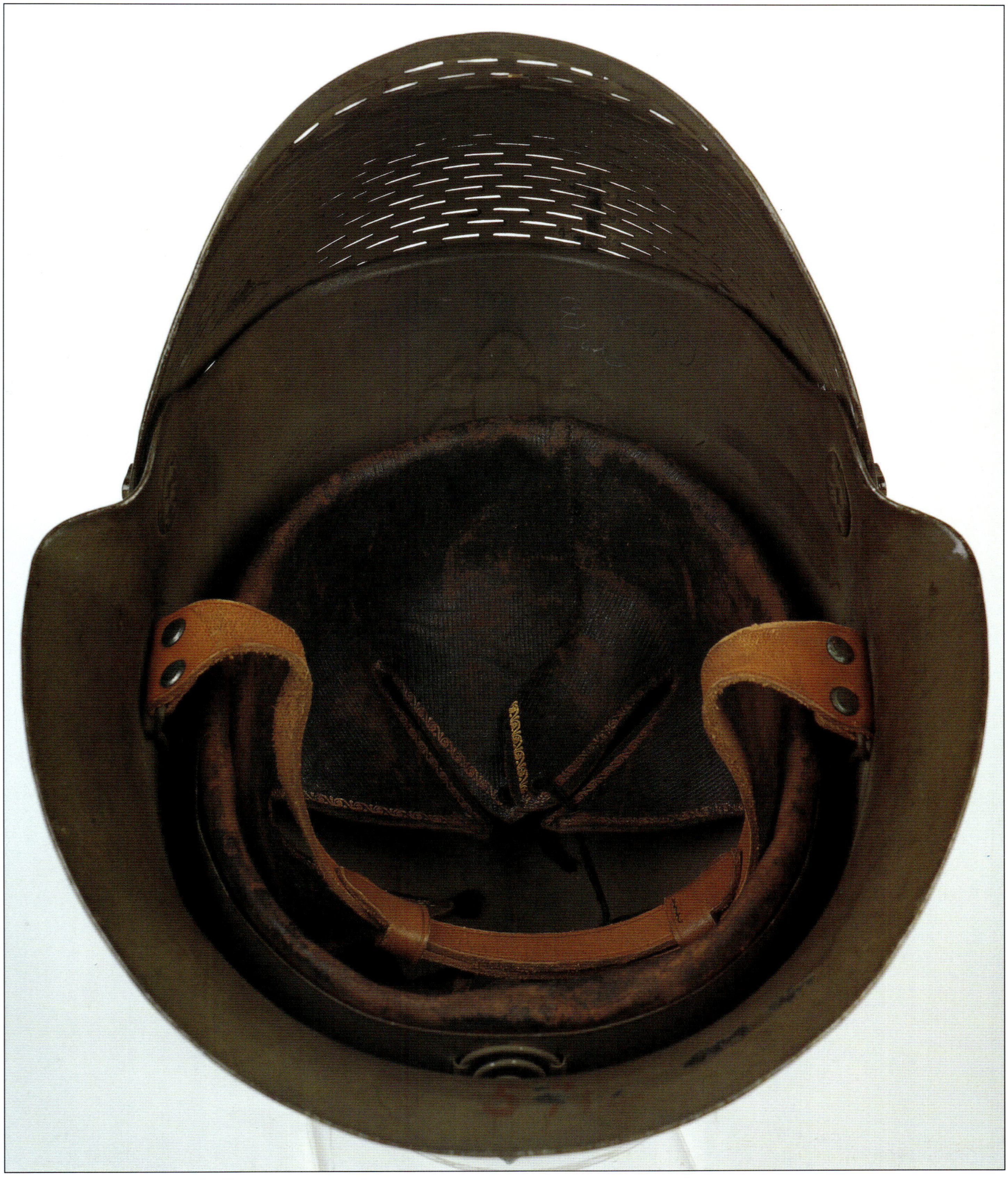

Experimental Helmet Model No.2 "Deep Salade". Note the use of the woven chinstrap as found on the later Model 1917A1 Helmet. (Collection of Marcus Cotton)

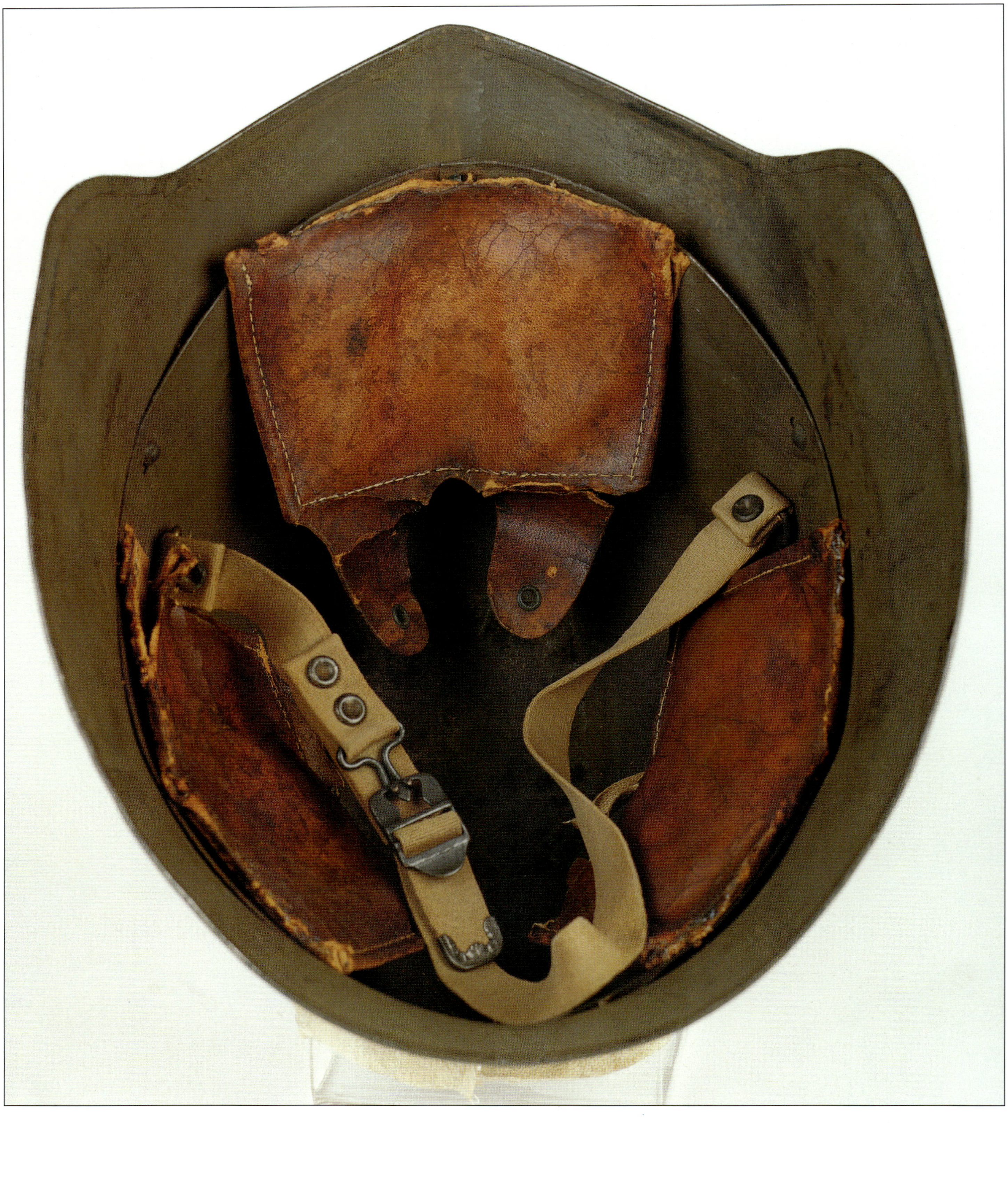

Experimental Helmet Model No.5. Note that this example has a leather chinstrap rather than the woven type. There were several other versions of the No.5 helmet, designated Experimental Model No.5A which had variations of helmet shell contours such as a more gentle slope between the rear brim and visor. (Collection of Marcus Cotton)

Experimental Helmet Model No.5. This example is camouflaged in the German style, probably as a souvenir. It also carries the "Keystone" insignia of the 28th Division. It is therefore quite possible that this example saw service in France. (Collection of Marcus Cotton)

Experimental Helmet Model No.7. (Sentinels' helmet) Note that the liner is missing in this example. These helmets were manufactured by William H. Mullins Co. of Salem Massachusetts USA. (Collection of the Imperial War Museum, London)

Experimental Helmet Model No.8. (Collection of Marcus Cotton)

This page and following pages. Experimental Helmet Model No.8. (Collection of Ellwood von Seibold)

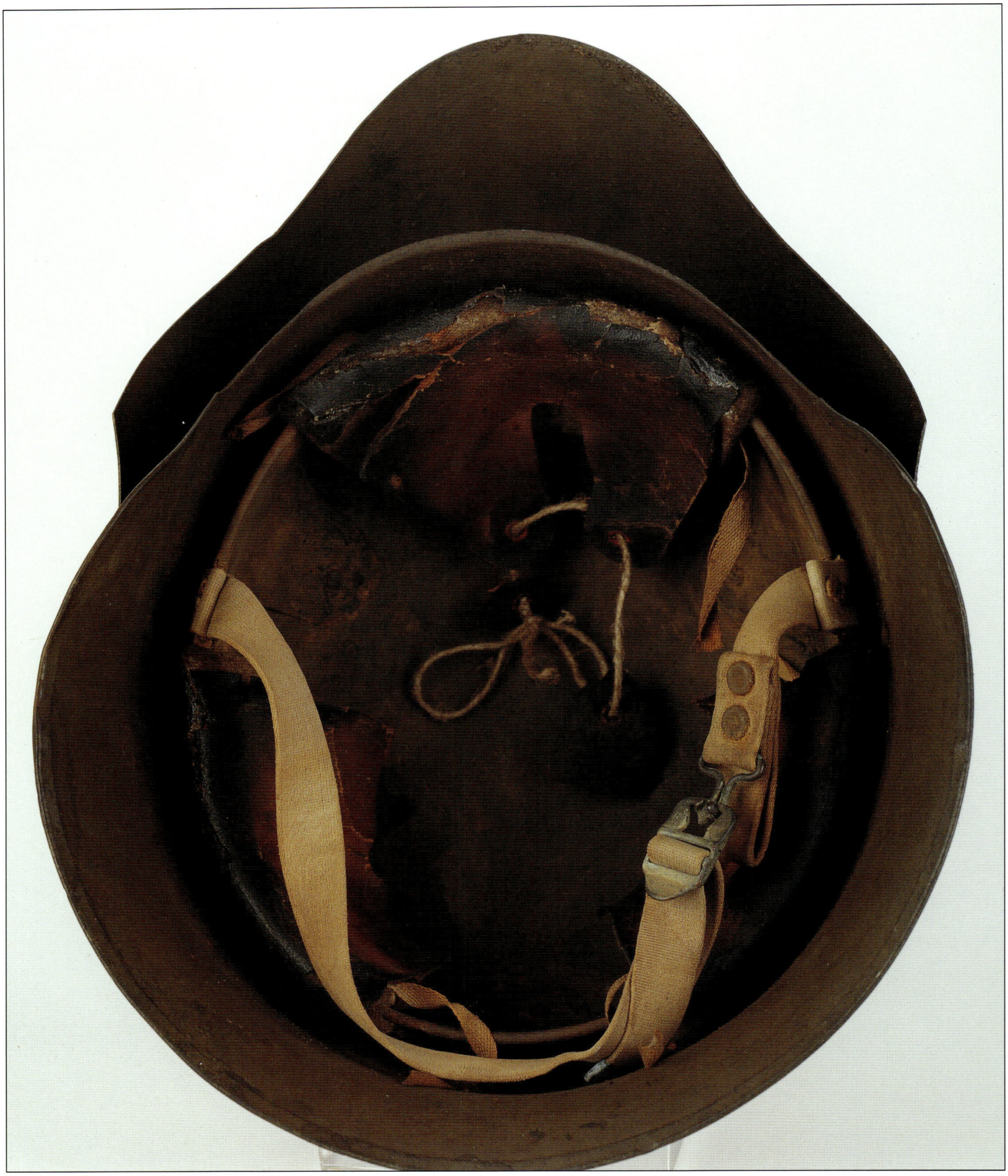

This page and following pages. The Liberty Bell Helmet. (Collection of Marcus Cotton)

Brigadier General James G. Harbold (C.O. Marine Brigade Chateau-Thierry). He is wearing a French-supplied Adrian Model 1915 helmet without badge fixing slots but with U.S. General's rank insignia fixed.

- Top: US troops wearing Adrian Model 1915 helmets.
- Bottom: American negro Regiments in the trenches. Possibly Maffrecourt 5 May 1918. Note they wear the French-supplied Adrian Model 1915 helmet. IWM Q69943

It is not possible to say with any certainty whether this soldier wears the British-supplied Mark I steel helmet or the American Model 1917 helmet.

■ Top: M.G. outpost. Amphersbach, Alsace. 29.8.18. IWM Q57009.

■ Bottom: Lewis gunners firing from a trench. Watau, 9th July 1918. IWM Q64336

A happy U.S. trenchman smiling. IWM HU 81573

■ Top: American engineers repairing broken wire during a gas attack: Andernach, 30th May 1919. This photo was taken during a post-war training exercise. IWM Q60987

■ Bottom: American soldiers wearing latest type gas masks. IWM Q60962

Comparison photograph of the Experimental Helmet No.2 "Deep Salade" with embossed national insignia to the front and the American Model 1917 helmet.

American soldier wearing steel helmet with visor and body armour: Boucq, 12.6.1918. The so-called visor is in fact the Wilmer Eye Shield. IWM Q80359

Result of bullet test on body armour. Langres, 27th April 1918.

Details from page opposite, left to right: Dunand helmet with Polack visor, Dunand helmet with Dunand visor, Adrian helmet with Polack visor. IWM Q112258

A soldier wearing a Dunand helmet with a Polack visor attached. U.S. National Archives.

The war is over! U.S. troops cheer as the ceasefire begins. U.S. National Archives

SOME BONNET
US
WALL
I EAT & DRINK OUT OF IT
IT PROTECTS MY BRAINS
IT'LL COME HOME WITH
ME UNDER IT

This page and following pages. Examples of steel helmets appearing in artwork for patriotic cards, posters, books and sculpture.

LA BAÏONNETTE
LES CAMOUFLEURS
peints par eux-mêmes

Die beste Sparkasse:
Kriegsanleihe!

Sommekämpfer
v. Kriegs-Internierten
Karl Volk.

Helft uns siegen!
Erler

Ich
wache und kämpfe für Dich
zeichne Du
Kriegsanleihe

LUDWIG
HOHLWEIN
MÜNCHEN

Und setzet ihr nicht das Leben ein / nie
wird euch das Leben gewonnen sein
FR. V. SCHILLER

That tin hat feels something like this on the way to the Offensive
And about like this when you get there
Bruce Bairnsfather

La toilette de Mams'elle Victoire.
The fashion for the Victory.

Still Keeping His Hand In
Private Smith (late Shinio, the popular juggler) appreciably lowers the protective value of his section's shrapnel helmets by practising his celebrated plate and basin spinning act.

NANNI
Ricompensa

Come On!
buy more
LIBERTY BONDS

Mauzan
...Sottoscrivete!

Note the American practice of splitting the chinstrap to stabilise the helmet.

Note the use of a cloth helmet cover on an American soldier's helmet.

BIBLIOGRAPHY

Gen. Franco Angioni
STORIA DELL' ELMETTO ITALIANO
RARA, Milano 1991

Chris Armold
STEEL POTS
R.James Bender Publishing
San Jose, 1997

Chris Armold
STEEL POTS VOLUME II. PAINTED STEEL
R.James Bender Publishing
San Jose, 2000

Ludwig Baer
THE HISTORY OF THE GERMAN STEEL HELMET 1916-1945
R. James Bender Publishing, San Jose 1985

J.A. Bowman
THE PICKELHAUBE VOLUME 1
Imperial Publications, Lancaster 1989

J.A. Bowman
THE PICKELHAUBE VOLUME 2
Imperial Publications, Lancaster 1992

Bruyelle, Mansuy, Raquillet
LES CASQUES DEL' ARMÉE FRANCAISE DE 1915 À 1945
ÉDITIONS E.S.P., Villette de Vienne

Stephen Bull
WORLD WAR ONE: BRITISH ARMY
Brassey's, London-Washington 1998

Stephen Bull
WORLD WAR ONE: GERMAN ARMY
Brassey's, London 2000

BULLETIN OFFICIEL DU MINISTÈRE DE LA GUERRE.
EDITION METHODIQUE. DESCRIPTION DES UNIFORMES
Charles Lavauzelle & Cie, Paris 1923

Stephen Chambers
UNIFORMS & EQUIPMENT OF THE BRITISH ARMY IN WORLD WAR 1
Schiffer Publishing Limited, Atglen PA 2005-06-26

Mike Chappell
BRITISH BATTLE INSIGNIA : 1914-1918
Osprey Publishing Limited, London 1916

Bashford Dean
HELMETS AND BODY ARMOR IN MODERN WARFARE
Carl J. Pugliese, New York 1919 & 1977

Martin Gilbert
FIRST WORLD WAR
Orion Publishing Group Limited, Londo 1994

Roland Hennequin
CASQUES MILITAIRES FRANCAISE DU XXE SIÈCLE
HISTOIRE ET IDENTIFICATION (1900-1944) TOME 1
Roland Hennequin, France 1998

Jacek Kijak
HELMY WOJSKA POLSKIEGO 1917-2000
Dom Wydawniczy Bellona, Warsaw 2004

Jürgen Kraus
THE GERMAN ARMY IN THE FIRST WORLD WAR
Verlag Militaria, Vienna 2004

LES CASQUE DE COMBAT DU MOND ENTIER DE 1915 À NOS JOURS TOMES 1, 2 & 3
Authors various
Charles-Lavauzelle, Paris-Limoges 1984

Emilio Lussu
SARDINIAN BRIGADE
Prion Books limited, London 2000

Francis Mackay
TOURING THE ITALIAN FRONT 1917-1919
Pen & Sword Books Limited, Barnsley 2002

Paolo Marzetti
ELMETTI
Ermanno Albertelli Editore, Parma 2003

Paolo Marzetti
LA GUERRA ITALO AUSTRIACA 1915-1919
Ermanno Albertelli Editore, Parma 1991

MILITARIA MAGAZINE
Histoire et Collections, Paris
Authors and articles various

Rest, Ortner & Ilming
THE EMPEROR'S COAT IN THE FIRST WORLD WAR
Verlag Militaria, Vienna 2003

THE TIMES HISTORY OF THE WAR
The Times, London 1920

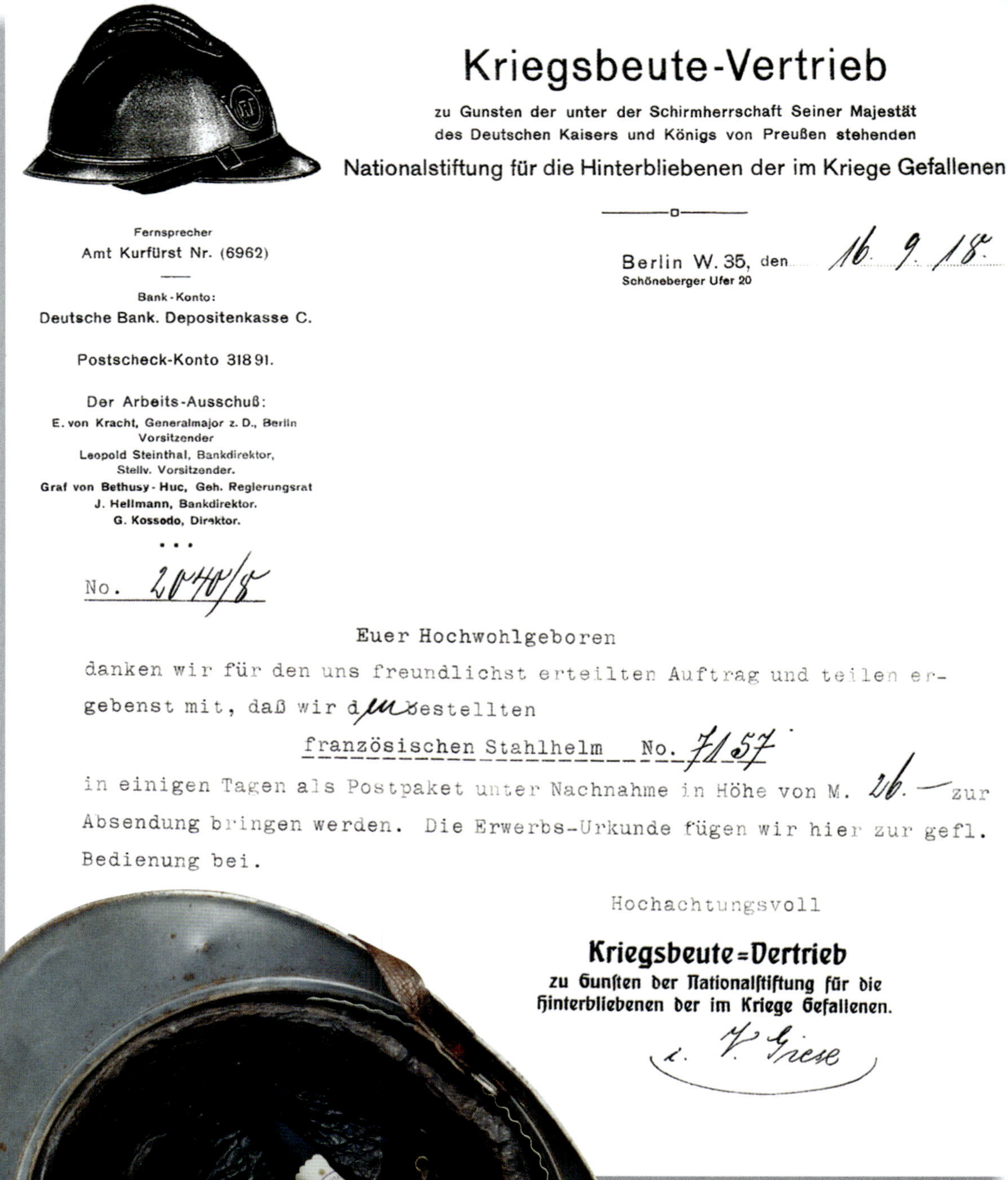

Kriegsbeute-Vertrieb

zu Gunsten der unter der Schirmherrschaft Seiner Majestät des Deutschen Kaisers und Königs von Preußen stehenden

Nationalstiftung für die Hinterbliebenen der im Kriege Gefallenen

Fernsprecher
Amt Kurfürst Nr. (6962)

Bank-Konto:
Deutsche Bank. Depositenkasse C.

Postscheck-Konto 31891.

Der Arbeits-Ausschuß:
E. von Kracht, Generalmajor z. D., Berlin
Vorsitzender
Leopold Steinthal, Bankdirektor,
Stellv. Vorsitzender.
Graf von Bethusy-Huc, Geh. Regierungsrat
J. Hellmann, Bankdirektor.
G. Kossodo, Direktor.

Berlin W. 35, den 16. 9. 18.
Schöneberger Ufer 20

No. 2040/8

Euer Hochwohlgeboren

danken wir für den uns freundlichst erteilten Auftrag und teilen ergebenst mit, daß wir den bestellten

französischen Stahlhelm No. 7157

in einigen Tagen als Postpaket unter Nachnahme in Höhe von M. 26.— zur Absendung bringen werden. Die Erwerbs-Urkunde fügen wir hier zur gefl. Bedienung bei.

Hochachtungsvoll

Kriegsbeute=Vertrieb
zu Gunsten der Nationalstiftung für die Hinterbliebenen der im Kriege Gefallenen.

i. V. Giese

It appears that in Germany captured helmets were sold in aid of charity. Shown is a copy of a letter dated 16.9.18 headed "War Booty Distribution" and confirming that French steel helmet No.7157 will be sent in the next few days under a collection order of 26 Marks in aid of The National Foundation of Surviving Families of Soldiers Who Died in the War. Although not the helmet referred to in the letter, a captured French Model 1915 helmet sold for charity is also shown. (Collection of Michael Haselgrove)